Cities and Design

Cities, initially a product of the manufacturing era, have been thoroughly remade in the image of consumer society. Competitive spending among affluent households has intensified the importance of style and design at every scale and design professions have grown in size and importance, reflecting distinctive geographies and locating disproportionately in cities most intimately connected with global systems of key business services. Meanwhile, many observers still believe good design can make positive contributions to people's lives.

Cities and Design explores the complex relationships between design and urban environments. It traces the intellectual roots of urban design, presents a critical appraisal of the imprint and effectiveness of design professions in shaping urban environments, examines the role of design in the material culture of contemporary cities, and explores the complex linkages among designers, producers and distributors in contemporary cities: for example fashion and graphic design in New York; architecture, fashion and publishing in London; furniture, industrial design, interior design and fashion in Milan; haute couture in Paris; and so on.

This book offers a distinctive social science perspective on the economic and cultural context of design in contemporary cities, presenting cities themselves as settings for design, design services and the 'affect' associated with design.

Paul L. Knox is University Distinguished Professor and Senior Fellow for International Advancement at Virginia Tech, where he was Dean of the College of Architecture and Urban Studies from 1997 to 2006. He has published widely on urban social geography and economic geography.

Routledge critical introductions to urbanism and the city

Edited by Malcolm Miles, University of Plymouth, UK
and John Rennie Short, University of Maryland, USA

International Advisory Board:

Franco Bianchini   Jane Rendell
Kim Dovey   Saskia Sassen
Stephen Graham   David Sibley
Tim Hall   Erik Swyngedouw
Phil Hubbard   Elizabeth Wilson
Peter Marcuse

The series is designed to allow undergraduate readers to make sense of, and find a critical way into, urbanism. It will:

- cover a broad range of themes
- introduce key ideas and sources
- allow the author to articulate her/his own position
- introduce complex arguments clearly and accessibly
- bridge disciplines, and theory and practice
- be affordable and well designed.

The series covers social, political, economic, cultural and spatial concerns. It will appeal to students in architecture, cultural studies, geography, popular culture, sociology, urban studies, urban planning. It will be transdisciplinary. Firmly situated in the present, it also introduces material from the cities of modernity and postmodernity.

Published:

Cities and Consumption – Mark Jayne
Cities and Cultures – Malcolm Miles
Cities and Nature – Lisa Benton-Short and John Rennie Short
Cities and Economies – Yeong-Hyun Kim and John Rennie Short
Cities and Cinema – Barbara Mennel
Cities and Gender – Helen Jarvis, Jonathan Cloke and Paula Kantor
Cities and Design – Paul L. Knox

Forthcoming:

Cities, Politics and Power – Simon Parker
Children, Youth and the City – Kathrin Hörshelmann and Lorraine van Blerk
Cities and Sexualities – Phil Hubbard

Cities and
Design

By Paul L. Knox

Routledge
Taylor & Francis Group

LONDON AND NEW YORK

First published 2011
by Routledge
2 Park Square, Milton Park, Abingdon, Oxon, OX14 4RN

Simultaneously published in the USA and Canada
by Routledge
270 Madison Avenue, New York, NY 10016

Routledge is an imprint of the Taylor & Francis Group, an informa business

© 2011 Paul L. Knox

Typeset in Times New Roman by
Keystroke, Tettenhall, Wolverhampton
Printed and bound in Great Britain by
TJ International Ltd, Padstow, Cornwall

British Library Cataloguing in Publication Data
A catalogue record for this book is available from the British Library

Library of Congress Cataloging-in-Publication Data
Knox, Paul L.
Cities and design / by Paul L. Knox.
p. cm.
1. City planning. 2. Sustainable development. I. Title.
HT166.K588 2011
307.1'216–dc22
2010001122

ISBN 13: 978–0–415–49288–1 (hbk)
ISBN 13: 978–0–415–49289–8 (pbk)
ISBN 13: 978–0–203–84855–5 (ebk)

ISBN 10: 0–415–49288–2 (hbk)
ISBN 10: 0–415–49289–0 (pbk)
ISBN 10: 0–203–84855–1 (ebk)

Contents

Figures

Tables

Acknowledgements

The author and publisher thank the following for granting permission to reproduce the following material in this work:

Jack Davis for Figure 1.1 'The baths at Vals, Switzerland'.

Hulton-Deutsch Collection/Corbis for Figure 1.2 'Paris 1900'.

Stock Montage/Getty Images for Figure 1.3 'The White City, Chicago'.

The MIT Press Journals for Figure 2.1 'The culture of design'.

The MIT Press Journals for Figure 2.2 'The culture of design: value, circulation and social practice'.

Tom Fox/Dallas Morning News/Corbis for Figure 2.4 'The Central China Television building in Beijing'.

Corbis Historical Picture Library for Figure 3.2: Leicester Square, London, *c.* 1750.

Cartography Associates for Figure 3.4 'Edinburgh New Town in 1834'.

Time Life Pictures/Getty Images for Figure 3.6 'Port Sunlight, Cheshire, England'.

Hulton Archive/Getty Images for Figure 3.12 'The Boulevard des Italiens, Paris, *c.* 1890'.

Topical Press Agency/Getty Images for Figure 3.13 'Letchworth, England'.

Stefano Bianchetti/Corbis for Figure 3.14 'Paris Metro entrance'.

Joseph Sohm/Corbis for Figure 3.15 'The Chrysler Building, New York City'.

Charles & Josette Lenars/Corbis for Figure 4.2 'Tapiola, Finland'.

Owen Franken/Corbis for Figure 4.4 'The Arche de La Défense, Paris'.

Bettmann/Corbis for Figure 4.5 'The demolition of the Pruitt-Igoe project in St Louis, 1972'.

Thomas A. Heinz/Corbis for Figure 4.7 'The Unité d'Habitation, Marseilles'.

Mark Fiennes/Corbis for Figure 4.8 'The Lever Building, New York City'.

Gail Mooney/Corbis for Figure 5.2 'Pier 17 at South Street Seaport, New York City'.

Alan Schein Photography/Corbis for Figure 5.3 'CityPlace, West Palm Beach, Florida'.

Preston Mack/Getty Images for Figure 5.4 'Main Street, Celebration, Florida'.

Photographie Werner Huthmacher for Figure 5.5 'Kirchsteigfeld, Germany'.

Johann Jessen for Figure 5.6 'Horseshoe Square, Kirchsteigfeld'.

Tim Graham/Getty Images for Figure 5.7 'Poundbury, Dorset, England'.

Jean Brooks/Robert Harding World Imagery/Corbis for Figure 5.8 'Spinnaker Tower and Gunwharf Quays, Portsmouth'.

Jason Hawkes/Getty Images for Figure 5.10 'London Docklands'.

Angus Oborn/DK Limited/Corbis for Figure 5.11 'The Stata Center, Cambridge, Massachusetts'.

Werner Dieterich/Getty Images for Figure 6.1 'The urban night-time economy: Grafton Street, Dublin'.

Bob Krist/Corbis for Figure 6.3 'The Guggenheim Museum, Bilbao'.

Royal Press Nieboer/Corbis for Figure 6.5 'Norwegian National Opera and Ballet, Oslo'.

Don Klumpp/Getty Images for Figure 6.6 'Quadracci Pavilion, Milwaukee Art Museum'.

Christopher Furlong/Getty Images for Figure 6.9 'Liverpool, the European Union Capital of Culture in 2008'.

Peter Taylor for Figure 7.1 'Cluster zones of architecture and engineering offices in London'.

Blackwell/Wiley for Figure 7.3 'Major concentrations of architecture firms with global practices'.

Blackwell/Wiley for Figure 7.5 'Flows of value surrounding fashion week events'.

Stephen Cotton/artofthestate for Figure 8.1 'Sheldon Square, London'.

Birkhäuser Verlag AG and Scott Campbell for Figure 8.2 'The Three Es of Sustainability'.

Frederick Florin/AFP/Getty Images for Figure 8.4 'Vauban, an eco-neighbourhood near Freiburg, Germany'.

PART I
Introduction

Part I provides an introduction to understanding and interpreting design in the context of the political economy of cities. Design plays a central role in the circulation and accumulation of capital. It also has to be understood in the context of the cultural and ideological changes associated with the shift from modernization to modernity and Modernism. Urban form, meanwhile, has to be interpreted in the context of the relationships among these changes. Cities and design also have to be understood in the context of the shift from mass production and mass consumption to the competitive consumption and the aestheticization of everyday life associated with 'romantic capitalism', the 'dream economy', and the 'society of the spectacle'. Professional design cultures and design movements are introduced within the framework of these economic, cultural and urban changes, together with the role of cities as crucibles of creativity and design innovation.

1 Cities, design and urban life

This chapter introduces design as an important aspect of urban life, in terms of the contributions of design not only to the functionality and aesthetic appeal of things but also to the broader sweep of economic, social and cultural change. Design can challenge, modify or reinforce these changes. Because of the dominance of consumerism in contemporary societies, design is critical to the successful marketing of all sorts of products. Through design, people feel that they can construct their own identities and their class distinction through their environment and their patterns of consumption. Cities are crucial settings for both the production and consumption of design. This chapter introduces concepts associated with the rise of consumerism, the aestheticization of everyday life, and the semiotics of things, and describes the roles of design in relation to modernization, modernity and Modernism.

Design has become a central aspect of contemporary urban life. Design can make things not only more attractive but also more efficient and more profitable. It is deployed not only in the development and redevelopment of neighbourhoods, buildings and interior spaces but also in the production of every component of material culture. Indeed, the claims that can be made on behalf of design extend to every aspect of urban life. Design can make urban environments more legible and can assist people in wayfinding (Gibson 2009). It can help people with physical disabilities through codified 'universal' design (Herwig 2008). It can promote and ensure public health (Moudon 2005) and bring order and stability to otherwise complex, chaotic and volatile settings (Greed and Roberts 1998). It can make transportation and land use more efficient (Wright *et al.* 1997; Levy 2008). It can be deployed for the benefit of women (Rothschild 1999), children (Gleeson and Sipe 2006), elderly people and those with disabilities (Burton and Mitchell 2006), minority populations (Rishbeth 2001) and social diversity (Talen 2008). It can prevent crime, protect built heritage, foster a sense of place, engender community, encourage conviviality, contribute to sustainability and combat climate change. It can signal social status and lifestyle, reflect taste and spearhead cultural change. It can make places more appealing, buildings more striking, clothes more stylish and objects more efficient.

But other important aspects of design concern its wider economic and symbolic value and its roles in supporting and sustaining the political economy of urbanized capitalism (Knox 1984, 1987; Cuthbert 2006). Because design can make places and things more efficient, safer, more functional, more attractive and more desirable, it is a vital dimension of the exchange value of things and a key determinant of their marketability – whether a building, a subdivision, a dress or a lemon squeezer. Because design can embody ideals and signal values, it is potentially a potent element of the dynamics of the political economy of places and nations. Together, these wider economic and symbolic issues are arguably the most significant aspects of design in relation to cities and urban life; they will be the dominant themes of this book.

Design in economic and social context

'In order to make sense of design', observes Adrian Forty, 'we must recognise that its disguising, concealing and transforming powers have been essential to the progress of modern industrial societies' (Forty 2005: 13). Design has an unambiguous role in facilitating the circulation and accumulation of capital, helping to stimulate consumption through product differentiation aimed at particular market segments. 'Designer' as an adjective has come to connote prestige and desirability, while 'designer' as a noun has to connote celebrity. Because of the prestige and mystique socially accorded to creativity, design adds exchange value to products, conferring a presumption of quality even though, like the emperor's clothes, this quality may not be apparent to every observer. Design also plays key roles in social reproduction, in the legitimation of authority, in the creation and maintenance of national identity, and in the absorption and deflection of ideas and movements that are potentially antithetical to dominant values and interests.

Most design historians recognize design as a specialist activity that emerged with the industrial revolution, mass production manufacture and consumer society. Yet, as John Walker observes, 'There appears to be a deeply-entrenched conservatism among design historians, an unwillingness to confront the relationship between design and politics, design and social injustice' (Walker 1989). Nevertheless, it is clear that,

> since design's beginning, when it was conceived as an art of giving form to products for mass production, it has been firmly embedded in consumer culture.

Design's first promoters in the 19th and early 20th centuries, Henry Cole in England and Herman Muthesius in Germany, for example, saw it exclusively in relation to the manufacture of products for the market.

(Margolin 1998: 83)

Kenneth Stowell, editor of *Architectural Forum* in the 1930s, acknowledged that 'architects . . . remain ultimately the highly paid employees of realtors and builders or are themselves small businessmen with a stake in the common exploitation' (quoted in F. Scott 2002: 47).

In the 1950s, the internationally respected designer George Nelson acknowledged that, by giving products a fashionable appearance, designers were virtually guaranteeing that they would seem obsolescent to consumers in a few years, thus continually stimulating demand and avoiding the market saturation. 'What we need', he added approvingly, 'is more obsolescence, not less' (Nelson 1956: 88). Since the 1950s, the underlying premise of design practice of all kinds – architecture, urban design and planning, interior design, product design, furniture design, fashion, photography, graphic design – has been that success ultimately depends on designers' sensitivity to the currents of trends and tastes within culture and on their ability to lend traction to capital accumulation by articulating these values and tastes to the promotion of ideas and events, services and products, buildings and cities.

Design, then, is a key instrument in the commodification and formatting of culture; it is fundamentally about styling, coding and effective communication with an audience of consumers. As William Saunders, editor of the *Harvard Design Magazine*, puts it with reference to architecture:

> along with every other cultural production (including music, photography, book publishing, the fine arts, and even education), the design of the built environment has been increasingly engulfed in and made subservient to the goals of the capitalist economy, more specifically the luring of consumers for the purpose of gaining their money.
>
> (Saunders 2005: vii)

Few are as unabashed about these roles as Kevin Kelley, whose architectural practice is advertised to clients as providing 'perception design'. His firm, he says, helps to

> prompt customers to buy through environmental 'signalling' that influences their perceptions. In a sense, we are designing the consumers themselves. Brand cueing takes place in the built elements but also the menu, uniforms, logo, aromas, and music plus sensations, and, most importantly, emotions. . . . We changed the firm's name to Shook with the tag line 'It's All Consuming'. We thus tell people that we eagerly embrace consumerism.
>
> (Kelley 2005: 53)

From a more general perspective, design can be seen as reflecting the *zeitgeist* of the prevailing political economy while serving, like other components of the system, as one of the means through which the necessary conditions for the continuation of the system are reproduced. Designers' roles as arbiters, creators and manipulators of aesthetics can be interpreted as part of the process whereby changing relationships within society at large become expressed in the 'superstructure' of ideas, institutions and objects. This allows us to see major shifts in design styles as dialectical responses to the evolving dynamics of urban-industrial society: part of a series of broad intellectual and artistic reactions to economic, social and cultural change. It also allows us to see design as a key instrument in the creation of national and metropolitan identities and the creation of class fractions and lifestyle groupings.

Another key role of design within the broader political economy is that of legitimation. Nineteenth-century businesses, for example, drew legitimacy from classical art, which had become closely associated with aristocratic and religious institutions. Hence department stores masqueraded as museums of art, banks were fitted out as ducal palaces, and factories were built to imitate castles. Today there is less imitation; instead, businesses acquire the originals – palazzi, stately homes and works of art – or sponsor museum spectaculars. A major theme in the literature on critical architectural history is the way that architecture has repeatedly veiled and obscured the realities of economic and social relations (Tafuri 1979). The physical arrangement and appearance of the built environment can help to suggest stability amid change (or vice versa), to create order amid uncertainty, and to make the social order appear natural and permanent. Thus there is a 'silent complicity' (Dovey 2000; Jones 2010) that exists between architects and the agendas of the politically and economically powerful.

Part of this effect is achieved through what political scientist Harold Lasswell (1979) called the 'signature of power'. It is manifest in two ways: through majestic displays of power in the scenography of urban design, and through a 'strategy of admiration', aimed at diverting the audience with spectacular and dramatic architecture. It must be recognized, however, that it may not always be desirable to flaunt power. Legitimation may, therefore, require modest or low-profile design. Conversely, it is by no means only 'high' design that legitimizes the prevailing order. The everyday settings of home, workplace and neighbourhood also help to naturalize class and gender relations. Thus another important function of design is in social reproduction, creating settings and images that structure and channel the values and world-views of different class fractions and that contribute to 'moral geographies' that express particular value systems in material form.

Design can also function to commodify critical or antithetical movements, thereby acting as an 'internal survival mechanism' of consumer capitalism and allowing the dominant social order to protect itself from opposing ideological forces. Through design, the energy of oppositional movements is diverted into commercialism, so that the movements themselves, having forfeited their raw power, pass quietly away. Think of the student and labour unrest of the late 1960s, for example, that challenged corporate capitalism and flirted with communal lifestyles, anarchism and revolution, only to be 'smothered beneath a cloying mass of Easy Rider posters and Love-and-Peace sew-on patches' (Knox and Cullen 1981: 184; see also Hebdige 1979; Frank 1998). Or think of the oppositional energy of the aggressive punk culture that sprang from dole queues and public housing estates in England in the late 1970s and early 1980s, only to be drained away as designers for boutiques, high street retailers and hair salons co-opted punk fashion, and as the strident and rebellious music of pioneer punk groups was drowned out by the catchier and more harmonious sounds of commercial post-punk bands.

More recently, the subversive and transgressive subcultures of rap and hip-hop have quickly been converted into mass markets for giant-sized T-shirts, low-slung baggy jeans and ostentatious jewellery. More substantially, as we shall see in Chapter 3, the radical oppositional impulses of nineteenth-century communitarian social reform movements were translated into professionalized urban design and planning that was charged with the management of urban settings as efficient places for business as well as healthy places for productive workers. And, to take just one more example – to be elaborated in Chapter 4 – the aesthetic of the seminal Modernism of the Bauhaus, originally tied closely to socialist ideals, was quickly co-opted by corporate capital when its leading practitioners crossed the Atlantic.

Design in contemporary society

Today, it seems, everything is designed, and a 'designer' aesthetic permeates almost every aspect of urban life:

> Few of the experiences we value at home, at leisure, in the city or the mall are free of its alchemical touch. We have absorbed design so deeply into ourselves that we no longer recognise the myriad ways in which it prompts, cajoles, disturbs an excites us. It's completely natural. It's just the way things are.
>
> (Poynor 2007: 136)

The reason, of course, is that Western economies have been based for decades on a culture of materialism that has incrementally ramped up the importance of style, fashionability and cool. Economic historians point to the 1920s as the moment when consumers' purchasing power began to match their aspirations, the mass production and mass consumption logic of Fordism unleashing a new socio-cultural phenomenon: competitive consumption. This was also the moment when the idea of the home as a privileged consumer durable became established, with private homes as the stage for materialistic lifestyles and the containers for an extended range of material possessions. In the economic boom after the Second World War, material consumption took on a more expansive form as discretionary spending by the middle classes reached unprecedented levels.

Harvard economist James Duesenberry (1949) identified the trend at an early stage, contrasting it with the nineteenth-century version of conspicuous consumption that had been documented by Thorstein Veblen (1899). Instead of being driven by an elite 'leisure class', postwar consumption was a middle-class suburban phenomenon, driven by neighbours: the eponymous 'Joneses'. Historian Lizabeth Cohen (2003) writes of the emergence in the 1950s of a 'consumers' republic' in the United States, based on the mass consumption of motor cars, houses and manufactured household goods, all celebrated by the new medium of television. Western Europe, recovering from the Second World War, lagged a decade or so behind.

Sociologist Colin Campbell (1987) writes of a 'spirit of modern consumerism' that had its origins at this time as people's lives became infused with illusions, daydreams and fantasies about consumer objects. Under the spirit of modern consumerism, people have come to constantly seek pleasure, enchanted by a succession of objects and ideas, always believing that the next one would be more gratifying than the previous one. This is the basis of what Campbell (1987) calls 'romantic capitalism', driven by the 'self-illusory hedonism' of dreams, fantasies and competitive consumption. Romantic capitalism was soon boosted by the widespread availability of credit cards. By the late 1960s, Guy Debord (1967: 42) had identified the emergence in Western culture of a *Society of the Spectacle*, defined as the 'moment when the commodity has attained the total occupation of social life'. Jean Baudrillard (1968: 24) wrote of 'the need to need, the desire to desire'. Baby boomers were coming of age and transforming the norms of consumption as well as politics and popular culture.

The formative experience of the baby boomers was the postwar economic boom. Growing up in affluent sitcom suburbs, they initially rebelled against the apparent complacency of what J.K. Galbraith (1976) had dubbed the 'Affluent Society', channelling their energies into countercultural movements, many of them with a vaguely collectivist approach to the exploration of freedom and self-realization. But in 1973 the quadrupling of crude oil prices by the Organization of the Petroleum Exporting Countries caused a global economic system-shock that sobered the boomers into a more materialistic and self-oriented world-view. Yet the economic circumstances of the 1970s did not permit a smooth transition to materialism.

The result was that people began to save less, borrow more, defer parenthood, comfort themselves with the luxuries that were marketed as symbols of style and distinctiveness, and generally surrender to the hedonism of lives infused with extravagant details: designer accessories, designer clothes, designer decor, designer fittings and furniture, and, for those who could afford it, fancy cars and designer homes in landscaped neighbourhoods. In the United States, the cultural hearth of hedonistic and competitive consumption, it added up to what conservative commentator David Brooks (2004) calls a 'Paradise Spell' of relentless individual aspiration and restless consumption, 'the controlling ideology of American life'.

The aestheticization of everyday life

By the 1980s, traditional identity groups based on class, ethnicity and age had begun to blur as people found themselves increasingly free to construct their identities and lifestyles through their patterns of consumption. In addition to the traditional business of positional consumption, members of new class fractions and affective 'neotribal' groupings sought to establish their distinctiveness through individualized patterns of consumption (Featherstone 1991; Bocock 1993; Maffesoli 1996). Thanks to the successes of Fordism, consumers' dreams could be fulfilled more quickly and more easily. Enchantment initially sprang from the affordability and choice resulting from rationalization and mass production. But this led inevitably and dialectically to disenchantment as novelty, exclusivity, distinction and the romantic appeal of goods were undermined by mass consumption. To counter this tendency, product design and niche marketing, along with the 'poetics' of branding, have become central to the enchantment and re-enchantment of things (Paterson 2006; Donald *et al.* 2009).

As sociologist George Ritzer (2005) has pointed out, enchantment also came to be ensured through a variety of specialized urban settings – 'cathedrals of consumption' – geared to the propagation and facilitation of consumption: shopping malls, chain stores, franchises and fast food restaurants, casinos and themed restaurants. Meanwhile, as mass markets became saturated, the mid-1980s marked the emergence of specialized consumer market segments, identified through market research by way of psychographics. Advertisements, playing to the sensibilities and dispositions of the Paradise Spell, consequently shifted away from the simple iconology of mid-twentieth century campaigns (presenting products as embodiments of effectiveness and quality) to exploit narcissism (portraying products as instruments of self-awareness and self-actualization), totemism (portraying products as emblems of group status and stylishness) and covetousness (baldly presenting products as emblems of exclusivity and sheer wealth).

The result was the aestheticization of everyday life, with design implicated in production and consumption at every level. The design of the built environment has become intimately involved with many aspects of consumption, especially those involving an explicit design premium, such as fashion and luxury products (Patton *et al.* 2004). Endorsement by association, observes Martin Pawley (2000),

> is one of the things that architecture does best, and also one of the things that fashion, the industry, needs most – the new car parked outside the manor house, the classical revival office building, the corporate headquarters campus, the view from the castle, the minimalist interior . . . All of them can be borrowed . . . to make or remake a reputation.
>
> (Pawley 2000: 7)

In other words, fashion and architecture use one another, not simply as backdrops or as ecologies for celebrity-laden events, but as guarantees of cultural accept-ability. The spa designed by Peter Zumthor in Vals, Switzerland, for example (Figure 1.1), has been used as a backdrop for fashion shoots, music videos and advertising in order to create a rarefied atmosphere and at the same time to appeal to a certain target group with architectural knowledge. High-end architecture and high-end fashion also have an affinity for one another because both require great precision in fabrication and construction, high levels of finish quality and carefully controlled lighting. Commodified, the relationship has produced a distinctive luxo-minimalism in interior design, with celebrity architects like Massimiliano Fuksas, Rem Koolhaas and John Pawson furnishing minimalist backgrounds for con-temporary fashion brands like Armani, Boss, Jigsaw, Calvin Klein, Mango, Issey Miyake, Prada and Louis Vuitton. A good illustration of the way that luxury goods producers seek to create a 'brand universe' for consumers through art and architecture is the luxurious coffee-table book *Louis Vuitton: Art, Fashion and Architecture* (Gasparina *et al.* 2009) that features the firm's collaborations with, among others, Frank Gehry, Zaha Hadid, Hans Hemmert, Anouska Hempel, Peter Marino and Richard Prince.

This is part of the emergence of a new interdependence among fashion, retail and architecture that has been prompted in part by the acquisition of elite couture houses by retail conglomerates, which quickly realized architecture's marketing and branding potential. 'Name' architects have been drawn increasingly into product lines – Michael Graves' kitchenware design for Target stores, Aldo Rossi's kitchen and table ware for Alessi, Mario Botta's Caran D'Ache fountain pen (retail: $2,100), Norman Foster's desk accessories for Helit, and so on – while couture houses have exploited their brand identity to sell everything from jeans and

Figure 1.1 The baths at Vals, Switzerland. Designed by Peter Zumthor and built between 1993 and 1996, the thermal baths have become a 'canonized commodity' after numerous media features in association with fashion and popular culture. The July 1997 issue of *Vogue*, for example, contained a ten-page swimwear feature called 'Body Building', each page consisting of a single photograph showing a model inside the building. Also in 1997, the video for Janet Jackson's song 'Every Time' from her album 'The Velvet Rope' was filmed in the baths. (Photo: A.J. Davis)

underwear to sunglasses and watches. In larger metropolitan centres, fashion retailing has also developed a synergy with commercial art galleries and public museums and galleries, emulating museum and gallery design in their stores (and sometimes even incorporating mini-exhibitions in their stores) and shadowing their geographical location in the city. As Lees (2001) notes:

> In exploring how architectural spaces are inhabited and consumed, geographers of architecture might take a page from developments in the new consumption literature. Geographers now argue that consumption should be seen as a productive activity through which social relations and identities are forged. Such a perspective on consumption as an active, embodied and productive *practice* dispels the sharp production/consumption distinction, and with it those tired debates about resistance to versus domination by the inauthentic consumerism of more or less duped consumers. The new geography of consumption recognizes 'the creativity of "ordinary consumers" in actively shaping the meanings of the goods they consume in various local settings', while insisting also that the commodities themselves, the processes of their production and the identities of their consumers cannot be thought of as fixed and essential but instead must be theorized as what Harvey calls 'structured coherences', or what Latour calls 'actants' that emerge as such through networks of inter-related practices.
>
> (Lees 2001: 55)

The trend has been spread and intensified as internet shopping has prompted retailers to offer something different: not convenience or cost savings but a special experience. For the people who can afford it, *performing* consumption now plays a key role in the construction of distinct and fashionable identities. 'As the trend for shopping online increases, so the power of three-dimensional space in the form of a retail outlet created as a sensory experience for the shopper increases in importance for the powerhouse brands' (Mackereth 2000: 61).

In mass markets, meanwhile, the 'corporate Cool Machine' (Frank 1998) closely monitors incipient consumer trends, youth cultures and countercultures, and commissions designers to generate and form products to profit from the trends. Consultants employ teams of young professionals to go undercover, monitor their peers and discover what's cool. The London 'guerrilla advertising' outfit Cake, for example, maintains a list of what it believes to be the 1,000 coolest people and companies. They mail their clients' new products to early adopters and ask them to fill in a questionnaire: are the products cool or not? Rick Poynor (2007: 75) suggests that 'cool' has become the dominant sensibility of advanced consumer capitalism: 'Cool wards off social embarrassment and offers a new (ironic) form of certainty. If you own cool things, then you too must be cool, since you are what you buy.' Cool, of course, is beyond words, like many other aspects of taste and aesthetic judgement. If you get it, no explanation is necessary; if you don't, no explanation is possible.

The semiotics of things

This points to the central importance of design in contemporary culture, facilitating the ways in which we are able to establish shared meanings and read off people's values, lifestyle and status from their possessions, the clothes they wear, and the landscapes they inhabit. Patterns of consumption are epigrammatic, able to carry sophisticated symbolic meaning. They mould people's consciousness of place and of each other, and help people to connect the realms of nature, social relations and individual identity. 'Surrounded by our things', writes McCracken (1988: 124), 'we are constantly instructed in who we are and what we aspire to'. Yet signification bears no straightforward relationship to the material world. Signs and symbols 'reflect and refract another reality. Social life is impregnated with signs which make it classifiable, intelligible, and meaningful' (Eyles 1987: 95). Each signifier, whether it is a house or a watch, a car or a pair of shoes, can be ascribed not only a denotative, surface-level meaning but also one or more second-level, connotative meanings. Within particular socio-cultural settings, certain signifiers are trans-formed – 'cooked', in the terminology of Lévi-Strauss (1970) – to form the basis of a socially constructed 'reality': a particular way of seeing the world.

Design alters the way people see commodities and geographic settings and establishes them as 'semiotic goods' whose economic value is based in part on the meanings people give them rather than their functionality. Such is the power of 'design' for its own sake that some goods can be successful in the marketplace in spite of having very little functionality at all. An oft-quoted example is the *Juicy Salif* lemon squeezer, designed for Alessi in 1990 by Philippe Starck and still selling, twenty years later, for more than $80 each. For reflexive consumers (i.e. those who continuously re-examine, reappraise and reconsider their consumption practices) who seek to build their identities through design objects, fashion and art, the fact that the *Juicy Salif* is practically unusable is offset by its cool looks and by the power of the combined brand identities of Alessi and Starck. Of course, if you've never heard of Alessi or Starck, the thing may just look strange. But, once you have acquired a certain amount of design knowledge, it is impossible to revert to a position of semiotic not-knowing.

Today, even modestly affluent households are sophisticated and reflexive, highly adept at the art of positional consumption. But the symbolism and meaning of material goods and the built environment is under constant construction and reconstruction, interpretation and reinterpretation by everyone, individually and collectively. New products, new designs and shifts in taste and style have the tendency to exclude those who may not be 'in the know' or do not have the means to make 'necessary' changes to their ensemble of possessions and patterns of activities.

At the level of the individual household, it is important to maintain a consistent aesthetic as new objects are incorporated and the old discarded. In marketing terms this is known as brand coherence. Its significance was recognized long ago by the French philosopher Denis Diderot, and it is sometimes referred to as the 'Diderot effect'. Diderot had been working quite happily in his crowded, chaotic and rather shabby study until he received a fancy velvet smoking jacket as a gift. He liked his new jacket but soon noticed that its quality made his surroundings seem threadbare. His desk, rug and chairs looked scruffy by comparison. So, one by one, he found himself replacing his furnishings with new ones that matched the jacket's elegant tone. He realized (though he later regretted it) that he had felt the need for a sense of coherence, a sense that nothing was out of place.

Consumers' design knowledge

Today's consumers, attentive both to brand coherence and to the subtle (and some-times sudden) shifts in the semiotics of things, select houses and purchase products, services and experiences that give shape, substance and character to their particular identities and lifestyles. Consumers' design knowledge comes from a variety of sources: advertisements, product placement in movies, television makeover programmes, print media, blogs and word-of-mouth, along with a great deal of tacit understanding that comes from social cues and people's reflexive awareness. The raw origins of much of this knowledge and understanding can be found in specialized, design-oriented print media of one sort or another: books, professional magazines, trade journals and niche-oriented lifestyle magazines. Books on design are overwhelmingly dominated by large-format coffee-table books and by monographs from publishers like Birkhäuser, Phaidon, Princeton Architectural Press, Rizzoli and Taschen that maintain specialized lists in architecture and design with a carefully cultivated sensitivity to the book-as-object. Design bookstores also stock the products of smaller specialized and vanity presses that publish the glossy body-of-work volumes with which architects and designers hope to impress both colleagues and future clients.

For the most part, the understanding of design derived from these books is decidedly narrow and usually framed around the persona of a particular designer, the products of a particular technology or technique, or the aesthetics of a particular category of objects. Coffee-table books, by definition, exempt the reader from intellectual effort; while the more specialized design literature, as Adrian Forty (2005: 6) observes, suffers 'from a form of cultural lobotomy which has left design connected only to the eye, and severed its connections to the brain and to the pocket'. The same fixation with aesthetics and personalities – design and designers, art and art directors, illustration and illustrators, photography and photographers – is evident in professional magazines, especially the so-called 'showcase' or 'portfolio' magazines such as *Abitare*, *Communication Arts*, *Domus*, *Graphis*, *I.D.* and *Print*, all of which are high-gloss productions that use sumptuous photography and printing techniques to show off the latest architecture, interior design, furniture design, graphic design, product design and packaging.

While this literature encompasses much of the formal, professionalized understanding of design, a much wider readership is attuned to what Sharon Zukin (1991, 2004) has described as the 'critical infrastructure' of consumption: consumer guides, the Sunday supplements, lifestyle magazines like *Architectural Digest*, *Elle Décor*, *Living*, *Metropolis*, *Metropolitan Home*, *Wallpaper**, *World of Interiors* and even 'magalogs' (hybridized, part magazine and part brand catalogue) like *Sony Style* and *A&F* [Abercrombie & Fitch] *Quarterly*. In these media, key cultural intermediaries – celebrities, editors, directors and copy writers – increasingly define what's cool and what's not. The layouts of lifestyle magazines make it perfectly clear that it is only in the combining of places – home, workplace, shop and recreation space – and in the juxtaposing of things – house, car, bicycle, shoes, bag, watch – that we fully articulate what we think we are. Discussions of furniture are blended with articles on clothing, architecture, industrial design and travel. As Deborah Leslie and Suzanne Reimer (2003b: 304) note, *Wallpaper**, a what-you-should-buy-next manual of what the editors present as 'urban modernism', has been the most prominent magazine within this genre. In *Wallpaper**, every item is captioned and priced. Its most striking innovation is the use of agency models to 'wear' the interiors and give them a normalized, 'lived in' look. 'The models become role models . . . showing us what we as occupants of these interiors should look like and how we should behave and dress' (Poynor 2007: 46–47).

Cities themselves receive similar treatment in the lifestyle magazines that are specific to particular metro regions. The modern city magazine movement was born in the United States in the 1960s, when most major metropolitan regions spawned publications bearing their name. Today there are more than 100 such publications in the United States. They have become part of the local critical infrastructure of consumption, vehicles for 'urban imagineers' who do not simply propagate a city 'brand' but also help to construct and impose sanitized and com-modified urban identities. Miriam Greenberg (2000) points out that while many of these city magazines started out with a broad coverage of local themes and issues, shifts in global, national and local dynamics base have forced cities to market themselves internationally in search of new sources of revenue:

> Through branding their city, these groups seek to forge emotional linkages between a commodified city and its increasingly footloose middle- and upper-class consumers (i.e., new potential residents, investors, corporate partners, tourists, and so on) in such a way that the name of the city alone will conjure up a whole series of images and emotions and with them an impression of value.
>
> (Greenberg 2000: 230)

The corporate publishers of these magazines have developed a common formula: toned down and reduced editorial content, increased advertising (commonly more than 60 per cent of content), coverage of consumption opportunities – restaurants, luxe malls and renovated waterfronts (rather than the city's people or natural or built environment) – as the brand identity of the city, and exhaustive high-end listings sections at the back.

Cities, modernity and design

Cities are engines of economic development and centres of cultural innovation, social transformation and political change. In the broadest of terms, we can identify four principal functions of cities in contemporary societies. First is the *mobilizing* function of cities. Urban settings, with their physical infrastructure and their large and diverse populations, are places where entrepreneurs can get things done. Cities provide efficient and effective environments for organizing labour, capital and raw materials, and for distributing finished products. Cities, in other words, are places where the classic economic advantages of centrality, agglomeration and what Alfred Marshall (1890) called 'industrial atmosphere' accrue to capitalist enterprise.

Second is the *decision-making* capacity of urban settings. Because cities bring together the decision-making machinery of public and private institutions and organizations, they come to be concentrations of political and economic power. Big cities, especially, are nodal command centres in the 'space of flows' that constitute contemporary space-economies.

Third are the *generative* functions of cities. The concentration of people in urban settings makes for much greater interaction and competition, which facilitates the generation of innovation, knowledge and information. Cities become, as Allen Scott (2001) puts it, 'creative fields'.

Finally there is the *transformative* capacity of cities. The size, density and variety of urban populations tends, as noted by nineteenth-century sociologists like Georg Simmel (1971) and Ferdinand Tönnies (1979), to have a liberating effect on people, allowing them to escape the rigidities of traditional, rural society and to participate in a variety of lifestyles and behaviours. More recently, Jane Jacobs (1969) pointed to the economic advantages enjoyed by cities as a result of their transformative and liberating capacity, arguing that high densities and socio-cultural diversity facilitates haphazard, serendipitous contact among people that, in turn, promotes creativity and innovation.

Design plays multiple roles in all of these dimensions of urbanization and urban life. It is central to the product differentiation at the heart of the consumer economy, implicated in the efficiency of urban settings as sites of production and consumption, subject to agglomeration (and therefore key to the nodality of some cities) and pivotal to many aspects of urban life at the intersection of economic, technological, social and cultural change. It is, in short, both a characteristic and a driver of contemporary urban life. Design is a crucial component of modernization, a product and a carrier of modernity, and a central tenet of Modernism.

Modernization and modernity

This, of course, requires some elaboration. All three – modernization, modernity and Modernism – have their roots in the seventeenth-century Enlightenment project that sought to advance reason, rationality and science over tradition, myth, superstition and religious absolutes. But they were fully unleashed by the twin revolutions of the late eighteenth century (the French Revolution) and early nineteenth century (the industrial revolution).

- *Modernization* refers to the processes of scientific, technological, industrial, economic and political innovation triggered by these revolutions and that also become urban, social and artistic in their impact (Berman 1983: 16–17).
- *Modernity* refers to the way that modernization infiltrates everyday life and permeates its sensibilities; the way that, as Baudelaire (1986) observed, urban life is characterized by the ephemeral, the fugitive, the contingent; and by speed, mobility, novelty and mutability.
- *Modernism*, meanwhile, refers to a wave of avant-garde artistic movements that, from the early twentieth century onwards, has responded in various ways to these changes in sensibility and experience.

Modernization had an immediate and direct effect on the significance of design. Specifically, the development of machine production made design very much more valuable to manufacturers. Adrian Forty (2005) gives this example:

> Maximizing the sales from each design had not been so crucial in handicraft industries, where, although profit might depend on the volume of production, there was not necessarily any advantage in using a single design rather than a variety of different ones. In the hand printing of calico, for example, additional output required more tables, more printers, and more blocks, but since each additional block had to be cut by hand, it made little difference if it was made to a new design or duplicated an existing one. The great advantage of machinery was its potential to manufacture a single design endlessly; the successful design became a very much more valuable possession, for *it was what released the machine's capacity to make a profit.*
>
> (Forty 2005: 58; emphasis added)

The discontinuities triggered by modernization – the unprecedented pace and scope of change, the way that time and space became abstract entities, the speed and power of new technologies, and the complexity of new social and institutional formations – meant that ambiguity, change and contradiction quickly became characteristic of modernity. Meanwhile, all aspects of life have become institutionalized, bureaucratized and commodified. Above all, modernity is dynamic, driven by two processes conceptualized by Anthony Giddens (1991) as disembedding and reflexivity. With modernity, traditional ways of doing things are disembedded and replaced by new ways in a process of continual change; while reflexivity means that there is a continual reappraisal and reconsideration of our lives in every sphere as we scrutinize guidebooks and magazines and consult experts and advisers. But, far from resulting in certainty, this leads to continually changing practices, trends and fashions. Experiment replaces tradition and popular culture develops a thirst for novelty. Progress is continually sought, yet constantly questioned and undermined. More paradoxically still, the seemingly unstoppable forward trajectory of modernization results in nostalgia – if not an overt longing for the past, then a formless regret and a melancholy feeling that something of the world has been lost. And this, in turn, feeds in to changing aesthetics and conceptions of beauty.

Professionalized design became an integral component of mass production and mass communication as a result of economic and technological modernization and the growth of middle- and working-class purchasing power. This shift was linked, as Guy Julier (2006) notes, to a 'visual turn' in Western society as the proliferation of images became commonplace and a key aspect of modernity:

> From a design point of view, commodities and services needed to be made more self-consciously visual in order to advertise and market them to a wide, anonymous audience. The Victorians saw the growth of the department store, catalogue shopping, mass tourism, and entertainment as spectacle – all of which hinge on the mediation of visual experience. And, of course, this also was the period of new visual technologies such as film, animation, and photography.
>
> (Julier 2006: 65)

As Robert Hughes points out in *The Shock of the New* (1980), the dislocations and new experiences introduced by modernization resulted in new ways of seeing and new ways of representing things. The places where all this was played out with the greatest intensity were the major cities of Europe. In London, the Pre-Raphaelite Brotherhood of painters, poets and critics set out to reform art, seeking to replace the reactionary classicism of the Victorian age. In Vienna, secessionists met in cafés in a ferment of new ideas about art, design, psychiatry and politics. In Zurich, Dadaists organized public gatherings and demonstrations, and established literary journals in the cause of anti-war politics and the destabilization of the prevailing standards in Western (high) culture; while in Milan, Futurists propagated the idea that the past was a corrupting influence on society, celebrating speed, technology and youth as the keys to the triumph of humanity over nature. But it was Paris that has come to be considered the capital of modernity, with its dramatic changes to the fabric of the city and to patterns of comportment and consumption (see Case Study 1.1).

Case Study 1.1 **Paris, capital of modernity**

Something very dramatic happened in Europe in general, and in Paris in particular, in 1848. . . . Before, there was an urban vision that at best could only tinker with the problems of a medieval urban infrastructure; then came Haussmann, who bludgeoned the city into modernity. Before, there were classicists, like Ingres and Davis, and the colorists, like Delacroix; and after, there were Courbet's realism and Manet's impressionism. Before, there were the Romantic poets and novelists (Lamartine, Hugo, Musset, and George Sand); and after came the taut, sparse, and fine-honed prose and poetry of Flaubert and Baudelaire. Before, there were dispersed manufacturing industries organized along artisanal lines; much of that then gave way to machinery and modern industry. Before, there were small stores along narrow, winding streets or in the arcades; and after came the vast sprawling department stores that spilled out onto the boulevards. Before, there was utopianism and romanticism; and after there was hard-headed man-agerialism and scientific socialism. Before, water-carrier was a major occupation; but by 1870 it had almost disappeared as piped water became available. In all of these respects – and more – 1848 seemed to be a decisive moment in which much that was new crystallized out of the old.

(Harvey 2003: 3)

Within a year of the Paris riots that led to political revolution in 1848, Louis-Napoléon Bonaparte (later declared Emperor Napoléon III) set about implementing much-discussed plans for urban renewal. A few years later, when Georges-Eugène Haussmann was assigned as prefect of the Seine département, the modernization of Paris gathered pace. He created wide boulevards, installed a new water supply system, a gigantic system of sewers and street (gas) lighting; built new bridges, a new opera house and other public buildings; laid out the enormous parkland of the Bois de Boulogne and made extensive improvements in smaller urban parks that turned them into places of sociality and leisure. Within this new framework, modernized industry flourished, along with significant new artistic and cultural movements, mass cultural entertainments and new spaces of consumption. It was no coincidence that the broad new roads meanwhile allowed for fast troop movement and crowd control. Haussmann had torn through the medieval urban fabric and carved up the city, peripheralizing the working class while offering vast opportunities to speculators.

Amid the turmoil of modernization Paris developed, as Harvey (2003: 223) puts it, a 'culture of governance and pacification by spectacle', and hosted a series of

Figure 1.2 **Paris 1900. Specially built pavilions and a giant globe surround the Eiffel Tower during the international Paris Exhibition of 1900. (Photo: Hulton-Deutsch Collection/Corbis)**

world expositions (1855, 1867, 1889 and 1900: see Figure 1.2). Amid the revolutionary ferment of ideas, Paris attracted and developed an unrivalled artistic and cultural scene that included, at various times, the artists Jean-Baptiste Corot, Pierre-Auguste Renoir, Gustave Courbet, Camille Pissarro, Henri de Toulouse Lautrec, Paul Cézanne, Vincent Van Gogh, Georges Braque, Pablo Picasso and Henri Matisse; the sculptor Auguste Rodin, philosopher Pierre-Joseph Proudhon, and writers Victor Hugo, Charles Baudelaire, Honoré de Balzac and Emile Zola. A radically reformed system of finance helped the government to put in place a modernized infrastructure, while new building technologies and new materials allowed for a spectacular increase in the scale of public and commercial buildings, none more so than the Palais de l'Industrie, built for the Universal Exposition of 1855.

The *passages couverts*, or arcades, of Paris were harbingers of the commodification and dazzling seduction of modern city life for those with the means to enjoy it. The first *passages* were built in Paris in the late eighteenth century by landlords who wanted to augment their income by exploiting the space within the blocks they owned. With its compact arrangement of diverse retail establishments, the *passage* was a new way to display and sell the mass-produced merchandise increasingly available in an age of industrialization. Protected from the weather by glass roofs, window shoppers were attracted by a variety of merchants, specialty stores and exhibitions (see Case Study 3.2). Glowing and magical at night with gaslight, the atmosphere in many *passages* changed, with the covered spaces offering shelter to strolling prostitutes and their customers.

Through the second half of the nineteenth century and into the first decade of the twentieth century, Paris also acquired a great modern market hall, Les Halles, of iron girders and skylight roofs; enormous new railway stations like the Gare du Nord and the Gare de Lyon; flamboyant new architecture and new landmarks like the Bibliothèque Sainte-Geneviève and St Augustin church; a mass transit system with Hector Guimard's famous wrought-iron Métro entrances; a proliferation of monuments, statuary, fountains and neatly tended parks; avenues of trees; new shops, department stores, restaurants and cafés; and the stupendous Eiffel Tower (completed 1889). The cityscape was eventually remade as a global object of desire and consumption, its aura enhanced by the exciting artistic and cultural life of the city.

Although Haussmann is widely credited with framing the setting for these mani-festations of modernity by creating unity out of the chaotic pre-modern city in a radical break with the past, David Harvey (2003) makes it clear that in Paris, as elsewhere, modernity was created by a slow process of modernization. The demolition of working-class quarters, the construction of elegant boulevards, the installation of streetlamps, the ordered uniformity of bourgeois apartment buildings, and the expansion of the city into the suburbs were only the material manifestations of a profound restructuring of economic and social relations prefigured by earlier changes in consumer culture, institutional frameworks and finance capital.

Key readings

David Harvey (2003) *Paris, Capital of Modernity*. London: Routledge.
Colin Jones (2005) *Paris: The Biography of a City*. New York: Viking.
Anthony Sutcliffe (1993) *Paris: An Architectural History*. New Haven, CT: Yale University Press.

The emergence of a new middle class and the growth of disposable income in the wake of the industrial revolution prompted German-born statistician Ernst Engel (1821–1896) to formulate what became known as Engel's Law: that as income levels rise, so households tend to spend proportionately more money on non-essential indulgences. Torstein Veblen (1899) famously wrote about the conspicuous consumption of the nouveau riche. More recently, Peter Dormer (1990) has identified the emergence of 'high design' for its own sake. He divides the concept into two categories: 'heavenly goods', combining high performance with exclusivity, that are designed for the rich to buy, and 'tokens', designed to be bought by the wish-they-were-rich.

Meanwhile, Georg Simmel (1971) had drawn attention to the role of fashion as an instrument of class differentiation within the relatively open and fast-changing society that had succeeded the old order. Fashion, like conspicuous consumption and high design, needs an audience, and here we should note the importance of the early antecedents of *Wallpaper**, *Metropolitan Home* and the like. Modern maga- zines began to emerge in the mid-nineteenth century, and by the end of the century a large number of mass-circulation titles had been established, covering art and architecture, interior design, fashion and women's consumer issues, along with trade journals for the building and furniture industries. These magazines were brought about by changing print and publishing technologies, and were made available on a national and international scale (Aynsley and Berry 2005).

Stimulated by mass communications, fashion has come to reflect and extend the thirst for novelty, innovation and the constant reinvention of the self that is so characteristic of modernity. Writing at the peak of the postwar Fordist boom in the late 1960s, sociologist Herbert Blumer (1969) acknowledged that fashion is a market-driven cycle of consumer desire and demand that operates as a means of class differentiation but argued that fashion is, at root, simply a response to people's desire to be in fashion, to be abreast of what has good standing, to express new tastes that are emerging in a changing world. A direct reflection, in other words, of the impulses and sensibility of modernity. Blumer's argument can be extended from dress to almost every object of consumption, especially now that many manufacturers seek profit not through mass markets (now close to saturation, thanks to the successes of Fordist mass production) but through niche markets, with products carefully designed to appeal to a particular lifestyle group or class fraction.

More recently, Elizabeth Wilson (2003) has explored the linkages between fashion and modernity in detail:

> In the modern city the new and different sounds the dissonance of reaction to what went before . . . The colliding dynamism, the thirst for change and the

heightened sensation that characterize the city societies particularly of modern industrial capitalism go to make up this 'modernity', and the hysteria and exaggeration well express it.

(Wilson 2003: 10)

Fashion, she observes, 'was always urban (urbane), became metropolitan and is now cosmopolitan, boiling all national and regional difference down into the distilled moment of glassy sophistication' (Wilson 2003: 9).

Wilson also writes of the way that fashion is a response to the dislocation and depersonalization of modernization, facilitating a sense of identity and bridging the relationship between the crowd and the individual, with the result a quintessentially modern ambiguity: that in dressing fashionably people simultaneously stand out and blend in:

> Fashion, then, is essential to the world of modernity, the world of spectacle and mass-communication. It is a kind of connective tissue of our cultural organism. And, although many individuals experience fashion as a form of bondage, as a punitive, compulsory way of falsely expressing an individuality that by its very gesture (in copying others) cancels itself out, the final twist to the contradiction that is fashion is that it often does successfully express the individual.

(Wilson 2003: 12)

Modernity begets Modernism

The flux of modernization presented a major intellectual challenge: how to understand the world, and how to represent it to ourselves. Artists and writers saw this as their special task, and amid the chaos of changing sensibilities felt liberated and empowered to explore every path, however radical. Thus we entered the age of the avant-garde, testing boundaries and challenging the status quo in the cause of social, political and economic reform; and sometimes simply pursuing art for art's sake, pushing the limits of aesthetic experience. By the early twentieth century, avant-garde design had coalesced around Modernism, a unifying system of ideology, technology and aesthetics that was seen by its advocates as a revolutionary stand against the elitism of nineteenth-century bourgeois culture.

Modernism was rooted in ideals of collectivism, standardization and social egalitarianism. Mass production and the invention of new materials were seen as enabling a democratization of material goods. Modernism thus embraced a belief in the progressive possibilities of new technologies and developed a rationalist philosophy expressed through aphorisms and declarations such as 'Form follows Function' and 'Less is More'. In architecture, Modernists came to believe that 'their new architecture and their new concepts of urban planning were expressing not just a new aesthetic image but the very substance of new social conditions which they were helping to create' (Carter 1979: 324).

Architecture was to be an agent of redemption. Through industrialized production, modern materials and functional design, architecture (as distinct from what Niklaus Pevsner dismissed as 'mere building') could be produced inexpensively, become available to all, and thus improve the physical, social, moral and aesthetic condition of cities. As Le Corbusier declared in his famous concluding lines of *Toward a New Architecture* (1927: 269): 'Architecture or revolution. Revolution can be avoided.' Interestingly, modern architecture and graphic design were already being deployed in the cause of revolution in the Soviet Union. Susan Buck-Morss (2002) has highlighted the parallels between the avant-gardes and utopian dreams of both Western capitalism and Soviet state socialism in the twentieth century.

In every area of design, Modernists wanted their designs to express the industrial age through a machine aesthetic that was underpinned by an ideology of scientific rationalism. They also subscribed to a moralistic aesthetic that held to 'integrity of surface' and 'truth to materials'. The corollary was an antagonism toward ornament and rapid stylistic change. As we shall see in Chapter 4, the full expression of Modernism and 'good design' came to fruition at the Staatliches Bauhaus, a school established in 1919 in Weimar, Germany, that combined architecture, crafts and the fine arts. Walter Gropius, the director of the Bauhaus from 1919 to 1927, rallied designers to 'the creation of type-forms for all practical commodities of everyday use' (Gropius 1926: 95).

The type-form was the idea of a perfect – or at least optimal – solution to the functionality of every product, every building. Modernist designers thus aimed to create goods that were immune to the logic of fashion cycles. It followed that the unsophisticated tastes of the untutored and bourgeois public needed to be reformed through the evangelistic discourses of the practitioners and advocates of Modernism. Modernism, observes Nigel Whiteley, 'supposedly rational, unsentimental, functional and serious – was about how architects and designers felt people *should* live; it did not grow out of the way people *do* live' (Whiteley 1993: 11; emphases in original).

As a result, there was an overwhelming emphasis on the creativity of individual designers, a cult of personality that quickly spread throughout the design professions, Modernist or otherwise. Nikolaus Pevsner's foundational account in 1936 of the history of design, for example (*Pioneers of the Modern Movement*), reads as the inevitable outcome of the work of inspired minds, a progressive shift toward a conclusion at the Bauhaus where the conflicts between art and industry are resolved. Sigfried Gideon sought to redress the balance a bit in his 1948 book, *Mechanization Takes Command: A Contribution to Anonymous History*, while another historian of design, Reyner Banham, hedged his bets, as indicated by the title of his 1960 book, *Theory and Design in the First Machine Age*.

But for the most part, the literature on design (and in design education) ever since has been to describe how 'the baton of genius or avant-garde innovation passes from the hand of one great designer to the next in an endless chain of achievement' (Walker 1989: 63). Not surprisingly, designed objects have become collectors' items, while successful designers have become celebrities. There is irony here in the consequent internal contradiction of the culture of Modernist design, whereby democratic intentions are displaced by elitism. But the greater irony is that the minimalist Modernist aesthetic, uprooted from its origins, quickly became commodified, the preferred form for corporate towers, luxury metro interiors and, most recently, high-end fashion retail settings.

Nevertheless, Modernism has by no means been hegemonic as the designed face of modernity. As we have noted, novelty and mutability are characteristic of modernization and modernity, and one outcome of this has been an interlude of 'postmodern' design. The prefix is misleading insofar as it implies an epochal shift away from modernization and modernity, whereas it can – and probably should – be seen as an important phase of modernity, driven by a combination of increasingly materialistic societies, widespread misgivings about the ability of modernization to deliver progressive economic and social outcomes, and increasing scepticism about the totalizing rationalism of Modernism. As we shall see in Chapter 5, postmodern design was a response to changing economic and social conditions, a facet of the 'society of the spectacle' in which the symbolic properties of places and material possessions assume unprecedented importance. Above all, postmodernism is consumption-oriented, with an emphasis on the possession of particular combinations of things and on the style of consumption.

Modernity and urban form: spaces of consumption

Modernization and modernity brought new kinds of structures, new patterns of urban form and, with them, new forms of sociability. In particular, the development of the modern city from 1880 to 1945, dominated increasingly by the proliferation of bourgeois commercial culture, saw the introduction of arcades, department stores, cafés, restaurants, tea rooms, dance halls, theatres, hotels, public parks, sports stadiums and amusement parks. Department stores and shopping arcades embodied the sense of flux, of kaleidoscopic motion, and of unceasing changes of images that many people found unsettling in modern cities. The new shopping and entertainment spaces used new materials and technologies – plate glass, cast iron and steel construction and coloured electric lights – to display their goods dramatically, especially at night.

Walter Benjamin, writing in the 1920s, noted how these consumption spaces – 'dream houses of the collective' – fostered the democratization of desire and a new urban culture based on acquisition and consumption and the cult of the new (Benjamin 1999; V. Schwartz 2001)). The new objects of desire, meanwhile, were periodically exhibited in 'spaces of triumph', World Expositions that were the ultimate halls of fantasy (Table 1.1). London's original Crystal Palace, built to house the 1851 Exhibition (the 'Great Exhibition of the Works of Industry of All Nations'), was the first spectacular urban megastructure. The World's Columbian Exposition in Chicago in 1893, accommodated in a 'White City' designed by Daniel Burnham and Frederick Law Olmsted (Figure 1.3), was explicitly intended to set a standard for architectural and urban design. Borrowing from European Beaux Arts models, Burnham's buildings helped the Fair 'to codify, by virtue of its "high culture" associations with Europe and Antiquity, the hierarchical framework by means of which "good taste" in America could be distinguished from "bad taste" ' (Rubin 1979: 344). Olmsted's overall plan of axes, basins, ponds and park-like landscaping helped to establish landscape itself as yet another commodity.

Table 1.1 World Expositions, 1851–1940

1851	London	1901	Buffalo
1855	Paris	1901	Charleston
1862	London	1904	St Louis
1867	Paris	1905	Liège
1873	Vienna	1906	Milan
1876	Philadelphia	1910	Brussels
1878	Paris	1911	Turin
1879	Sydney	1913	Ghent
1880	Melbourne	1914	Lyon
1884	New Orleans	1915	San Francisco
1888	Barcelona	1915	San Diego
1889	Paris	1929	Barcelona
1893	Chicago	1933	Chicago
1896	Nizhny Novgorod	1935	Brussels
1896	Budapest	1937	Paris
1897	Brussels, Stockholm	1939	New York
1900	Paris	1939–1940	San Francisco

Sources: Bureau International des Expositions (www.bie-paris.org/main/index.php?p=257&m2= 253) and ExpoMuseum (www.expomuseum.com)

Figure 1.3 The White City, Chicago. Built in Beaux Arts neoclassical style and designed by Daniel Burnham and Frederick Law Olmsted, the 'White City' was created for the World's Columbian Exposition in Chicago, in 1893. This photograph shows the massive Manufacturers and Liberal Arts Building. (Photo: Stock Montage/Getty Images)

In larger cities, these new spaces of consumption were drawn together by mutual attraction into distinctive new districts. In New York, for example, a new retail district emerged in the nineteenth century, stretching from Fifth Avenue to Union and Madison Squares and extending west to Sixth Avenue and east to Broadway. It was novel at the time in that it was almost exclusively oriented to retailing, with a mixture of large department stores and small boutiques interspersed with cafés and restaurants serving their customers and window-shoppers. Mona Domosh (1996) has shown how these spaces were designed to draw middle-class women into the city. In keeping with bourgeois taste, buildings were given elaborate decorative detail and an air of grandeur and civility that masked vulgar commercialization. In keeping with gender roles and sensibilities, the stores feminized their spaces with domestic motifs and floor plans.

For the most part, though, modernization and modernity brought about an increasing separation of public and private spaces. Doreen Massey (1991) notes that spatial and social reorganization meant that the city was increasingly gendered, with women increasingly relegated from the public to the private sphere of home and suburb:

The public city which is celebrated in the enthusiastic descriptions of the dawn of modernity was a city for men. The boulevards and cafés – and still more the bars and the brothels, were for men – the women who did go there were for male consumption.

(Massey 1991: 47)

Because 'good women' were to be kept from the gaze of men, the private realm became ever more private as the nineteenth century advanced, while the home became the locus of family consumption and the setting for the display of material goods.

The dream economy, design and consumption

It was at the onset of what Colin Campbell (1987) called 'romantic capitalism' that design became totally pervasive, an indispensable aspect of consumer society. Paul Glennie (1998: 928) urges us to note the distinction between consumption (relating to the volume and taxonomy of all spending that was in any way potentially variable or discretionary) and consumerism (relating to broader motivational drives to consume in particular ways). For Campbell, he notes, modern consumerism 'denoted general orientations to the accumulation of goods, the display of consumption and an unceasing search for novel experiences. At that time, consumption decisions by individuals were generally seen as a process of social positioning, whether explicit or implicit' (Glennie 1998: 928).

More recently, Glennie and others have emphasized the relationship between consumption and identity, through investigations of how meanings are created for, and transferred among, people, commodities, practices and social groups. Consumerism, from this perspective, 'is less a drive to possess goods as part of a process of social competition, than a drive to construct an independent identity through consumption activity' (Glennie 1998: 928). No longer is position ascribed by birth: rather, people are able to choose various types of identity through the goods they consume. This process has been described as the aestheticization of consumption. As Jackson and Thrift (1995: 227) note, 'identities are affirmed and contested through specific acts of consumption'. It is argued that in purchasing particular products and selecting particular house styles and neighbourhoods, people not only differentiate themselves from others but also find a means of self-expression in which they can adopt and experiment with new subject positions. Thus it is argued that people are increasingly defined by what they consume as well as by traditional factors such as their income, class or ethnic background. How we spend our money, in short, is now as important – or more so – than how we make it.

The resulting aestheticization of consumption is a far cry from past visions of twenty-first-century living inspired by Modernism. Virginia Postrel, in her book *The Substance of Style* (2003), describes the current fascination with mass customization and aesthetics. It is worth quoting her at some length:

> The twenty-first century isn't what the old movies imagined. We citizens of the future don't wear conformist jumpsuits, live in utilitarian high-rises, or get our food in pills. To the contrary, we are demanding and creating an enticing, stimulating, diverse, and beautiful world. We want our vacuum cleaners and mobile phones to sparkle, our bathroom faucets and desk accessories to express our personalities. We expect every strip mall and city block to offer designer coffee, several different cuisines, a copy shop with do-it-yourself graphic workstations, and a nail salon for manicures on demand. We demand trees in our parking lots, peaked roofs and decorative facades on our supermarkets, auto dealerships as swoopy and stylish as the cars they sell . . . To succeed, hard-nosed engineers, real estate developers, and MBAs must take aesthetic communication, and aesthetic pleasure, seriously. We, their customers, demand it.
>
> (Postrel 2003: 4)

Patrick Jordan (2007) has characterized the contemporary emphasis on pleasure, experience and aesthetics as the era of the Dream Economy:

> In the Dream Economy, success in the marketplace is not only about meeting people's practical needs, but also about meeting their aspirations and providing a positive emotional experience. . . . In the Dream Economy, consumers are increasingly looking for products and services which will meet their higher needs, which will enhance their self-image, express their values, enhance their relationships and perhaps even help them move towards self actualisation.
>
> (Jordan 2007: 6)

Commercial success in the Dream Economy, Jordan argues, depends on being able to design 'pleasurable products' that relate to trends in people's lifestyles and priorities. In the Dream Economy, what matters most is the way that goods and services are perceived to shape people's identity and mediate their social relations.

Identity and design

Social theorists have long grappled with this question of the relationship of consumer goods to social and cultural systems and to individuals' identities within them. Between the 1930s and the 1960s scholars of the Frankfurt School, led by Max Horkheimer, Theodor Adorno and Jürgen Habermas, drew on Marx's concepts of alienation and commodity fetishism in asserting that mass production, together with advertising, design and propaganda, amounted to a 'culture industry' that results in the pacification, coercion and manipulation of the masses (Bottomore 2003). According to this perspective, people's dreams are systematically shaped and appropriated for profit. The culture industry convinces people to trade off boredom and exploitation at work for leisure hours during which they can indulge in the pleasures of popular culture, shopping, watching movies and professional sport, and taking brief holidays. But this does not sit comfortably with the social and economic changes of the past several decades: the blurring of traditional identity groups based on class, ethnicity and age and the increasing tendency for new class fractions and affective 'neotribal' groupings to establish their distinctiveness through aestheticized, individualized patterns of consumption.

There is now a rich array of alternative concepts and theories in the literature on social change, consumption, identity, class, space, landscape and power. Drawing on Ferdinand de Saussure's foundational ideas on semiotics (Saussure *et al.* 2006) – the study of the meanings of signs and symbols and how they relate to the things or ideas they refer to – Jean Baudrillard (1981, 1998) has emphasized the importance of the symbolic component of consumption and its role in class differentiation and intra-class social rivalry. His analysis of codes of consumption has pointed to the importance of simulation – in themed shopping districts, festival settings and simulated communities, for example – in contemporary culture. Guy Debord (1993) has elaborated the overarching concept of the spectacle to describe consumer society and the ubiquitous packaging, promotion, display and media representation of commodities.

Others have pointed to the ways in which power and authority become stabilized and legitimized through codes of consumption and the symbolic content of landscape. Michel de Certeau (1984) emphasized the 'secret murmurings' of cues and codes embedded in the everyday experience of place and space: the underpinnings of inclusion and exclusion and the basis for people's negotiation of identity. Roland Barthes (1973) has explored the ways in which innocent-seeming social symbols form 'codes of domination' that sustain authority – sets of meta-signifiers that he called 'mythologies'. Michel Foucault (1997) pointed to the way that power and authority operates through 'normalizing regimes' of social and spatial practices. Pierre Bourdieu (1993) has explored the ways in which people's everyday experience results in a distinctive 'habitus' – a set of structured predispositions and ways of seeing the world.

Design, in context of these theoretical perspectives, plays a key role in the re-enchantment – or, as Guy Julier (2000) puts it, the 'de-alienation' – of products, settings and services as their novelty, exclusivity, distinction and romantic appeal are undermined by mass consumption, and as their meanings are appropriated by particular class fractions. In conjunction with marketing, branding and advertising, design meanwhile opens up the possibilities for individuals to attach sophisticated individualized meanings to things, and to construct their identities around them:

> Simultaneously or alternatively, the consumer builds his or her own hermeneutical engagement independently of the producer. The designer, in combination with other professionals, must take decisions regarding the degrees and mechanisms of de-alienation. The designer can entirely hand over the object to be received 'as it is', to allow the consumer to build their own meanings or none at all. Or he or she can carefully construct an 'aesthetic allusion' around the product.
>
> (Julier 2000: 55)

Design, then, should be seen as linking the economic to the cultural, as both articulating and enacting social relations and human behaviour; while designed objects, clothes, places and settings are subject to constant transformation and rewriting by both producing agents (designers, marketers and so on) and by their consumers and users.

Habitus and field

Members of particular market segments and class fractions develop collective perceptual and evaluative schemata – cognitive structures and dispositions – that derive from their everyday experience. These schemata operate at a subconscious level, through commonplace daily practices, dress codes, use of language and comportment, as well as patterns of consumption. The result is what French social theorist Pierre Bourdieu has conceptualized as 'habitus': a distinctive set of structured beliefs and dispositions 'in which each dimension of lifestyle symbolizes with others' (Bourdieu 1984: 173). Habitus incorporates both habit and habitat. It frames the sense of one's place in both the physical and social senses: it 'implies a "sense of one's place" but also a "sense of the other's place". For example, we say of an item of clothing, a piece of furniture or a book: "that's petit-bourgeois" or "that's intellectual" ' (Bourdieu 1990: 113). Or perhaps 'This is me', 'That's you' or 'That's cool'.

Habitus exists in different 'fields' of life. Bordieu's concept of field refers to subject areas (legal, political, scientific, artistic, academic, sociological, for example) with their own logic and with their own valuing of objects and practices. In every field, in other words, there are key signifiers that serve as a frame of reference for checking one another out. The key signifiers in the subject area of housing, for example, may be viewed as relating to questions of what constitutes a desirable house, how different rooms are used, how it should be furnished, how big it should be, etc. People's homes are clearly an important means of reaffirming and delineating class cultures – they are what Bourdieu (1990) referred to as 'structuring structures'. In this sense, housing represents a durable framework that serves to sustain lifestyles in accord with owners' dispositions. Other fields within which class fractions develop a distinctive habitus include food, art, music and fashion (Entwistle and Rocamora 2006). Habitus and field, then, are mutually interdependent, as a habitus always exists in relation to a given field. Fields are arenas of conflict and competition; habitus provides the players in each field with a sense of how to keep score.

Bourdieu's interest is in how culture is made to serve social functions, and specifically how culture is used to conceal the true nature of the power relations between groups and classes. According to Bourdieu, each class fraction will seek to sustain and extend its habitus (and new groups will seek to establish a habitus) by exercising its economic, social and cultural capital:

- Economic capital is any form of wealth that is easily turned into money.
- Social capital derives from the social connections of family, place and neighbourhood, often inherited through class membership.
- Cultural capital is the sum of a person's knowledge, skills and credentials. It can take three forms: *institutionalized* cultural capital, or formally accredited learning; *objectified* cultural capital, such as art, books and the stylistic aspects of architecture, interior design, product design and furniture; and *embodied* or 'symbolic' cultural capital, the non-accredited and sometimes tacit knowledge, tastes and dispositions absorbed through participation in a particular habitus. Embodied cultural capital derives from the command of superior taste. In traditional Western social hierarchies, superior taste is the distinguishing attribute of the habitus of bourgeois class fractions, whose members are socialized to appreciate fine art, music and 'good' design, to like certain foods, to understand complicated art forms, and to master certain context-specific manners, vocabularies and demeanours.

Different class fractions and market segments have different combinations of economic, social and cultural capital. Traditional upper-class groups command high levels of each kind of capital; working-class groups have little of any kind; the nouveau riche have plenty of economic capital but less social and cultural capital, and so on. The point here is that these combinations of different kinds of capital result in different dispositions: habitus. The fact that cultural capital is vulnerable to shifts in the denotative and connotative meaning of goods and practices and to the availability of new product lines and new designs only makes it more potent as a measure of distinction.

The habitus of dominant class fractions is inevitably undermined by the accessibility and popularization of goods and practices that were formerly exclusive, so that the process of maintaining habitus is continuous. The signs and symbols of distinction have constantly to be shuffled, inverted or displaced. New languages of taste and identity have to be mastered. New products and designs have to be inducted as desirable or enchanting (or not); the once-fashionable has to be condemned as dated or tasteless; kitsch has to be consecrated as cool; and showiness has to be cultivated as a desirable trait. In this context, the cultural capital available to a class fraction is critical. Cultural capital is not simply equivalent to cultural literacy but is, rather, a 'feel for the game', a product of knowledgeability of the symbolic meaning of particular cultural artefacts and socio-cultural practices.

As a result, dominant groups must continually pursue refinement and originality in their lifestyles and ensembles of material possessions. Less dominant groups, meanwhile, must find and legitimize alternative lifestyles, symbols and practices in order to achieve distinction. Subordinate groups are not necessarily left to construct a habitus that is a poor copy of others', however: they can – and often do – develop a habitus that embodies different values and 'rituals of resistance' in which the meaning of things is appropriated and transformed.

As Sharon Zukin (1991) notes, the increasingly complex and diverse world of art and design means that there is a growing need for cultural intermediaries – commentators, reviewers, editorialists, brokers, consultants, bloggers, copy writers, movie directors, talk show hosts – to intervene to aid in maintaining a 'feel for the game'. Designers themselves often create 'mood boards' in order to locate their task within what they perceive as the lifestyle (i.e. habitus) and knowledgeability (i.e. cultural capital) of their target market. Mood boards involve the arrangement and presentation of images of related buildings, landscapes, products and graphics onto blank poster boards in order to construct an associational context for whatever it is that is being (re)designed. Competing designs and people's behavioural traits are often added to mood boards, so that the product under design is also located as a marketing proposition.

Design, then, stands in the middle of a 'consuming paradox', (Miles 1998) wherein people feel that they can construct their own identities and their class distinction through consumption, while consumption simultaneously plays an ideological role in shaping and controlling the character of everyday life.

Space and society in contemporary cities

Today's urban and metropolitan regions are fragmented mosaics of commercial, residential and mixed uses that have overwritten the twentieth-century framework of sectors and zones focused on a dominant city centre. Downtown areas have been reconfigured as they have evolved from hosting the flagship shops and the headquarters offices of local firms to accommodating the regional offices of national and international firms. New industries associated with the 'new economy' based on digital technologies, biotechnology and advanced business services have sought new settings, for the most part well away from congested city centres where the built environment is ill suited to their needs.

Modern just-in-time production systems and flexible specialization strategies require easily accessible factories; biotechnology firms require specialized new laboratories; while almost every back-office facility and business service requires buildings that are flexible in layout and pre-wired or easily wired for access to digital communications networks. Business and industrial parks are based on single-storey structures with designer frontages, loading docks at the rear, and interior spaces that can be used for offices, research and development (R&D) labs, storage or manufacture, in any ratio. To be competitive, they must also be packaged as 'planned corporate environments' with built-in daycare facilities, fitness centres, jogging trails, restaurants and convenience stores, lavish interior decor and lush exterior landscaping and signage. Along with decentralized industry came retailing, service and office functions: strip malls, big-box discount stores, integrated shopping malls, specialized malls, power centres, fast-food franchises, hotel chains, family restaurants, corporate headquarters complexes, office parks, hopscotch sprawl and proliferating off-ramp subdivisions.

As a result, traditional concepts and labels – city, suburb, metropolis – are fast becoming examples of what sociologist Ulrich Beck calls 'zombie categories', concepts that embody nineteenth- to late-twentieth-century horizons of experience distilled into a priori and analytic categories that still mould our perceptions and sometimes blind us to the significance of contemporary change (Beck *et al*. 2003). Vast tracts of contemporary urbanized areas now consist of a 'metroburbia' of residential settings in suburban and exurban areas that are thoroughly interspersed with office employment and high-end retailing (Knox 2008). The social ecology of these settings is differentiated into neighbourhoods that reflect the segmented lifestyle groupings and class fractions of postindustrial society, each with different degrees of access to economic, social, cultural and symbolic capital.

These landscapes reflect the economic logic, social organization and cultural values of contemporary society. As Lewis Mumford (1938: 403) put it during an earlier phase of urban transformation: 'in the state of building at any period one may discover, in legible script, the complicated process and changes that are taking place within civilization itself'. Sociologist Ruth Glass (1968: 51) simply characterized the city as 'a mirror . . . of history, class structure and culture'. Kim Dovey (1999) puts it this way:

The built environment reflects the identities, differences, and struggles of gender, class, race, culture, and age. It shows the interests of people in empowerment and freedom, the interests of the state in social order, and the private corporate interest in stimulating consumption. Because architecture and urban design involve transformations in the ways we frame life, because design is the imagination and production of the future, the field cannot claim autonomy from the politics of social change.

(Dovey 1999: 1)

Within this broad framework, though, there is another important relationship between space and society. The city is never simply a straightforward mirror or neutral container of social processes. Particular urban spaces are created by specific sets of people, and they draw their distinctive character from the people that inhabit them. As social groups occupy urban spaces, they gradually impose themselves on their environment, modifying and adjusting it, as best they can, to suit their needs and express their values. Yet at the same time people themselves gradually accommodate both to their physical environment and to the people around them. There is thus a continuous two-way process, a socio-spatial dialectic (Soja 1980), in which people create and modify urban spaces while at the same time being conditioned in various ways by the spaces in which they live and work. Urban spaces, in the language of social theory, are both structured and structuring. As neighbourhoods and communities are created, maintained and modified, the values, attitudes and behaviour of their inhabitants meanwhile cannot help but be influenced by their surroundings and by the values, attitudes and behaviour of the people around them. At the same time, the ongoing processes of urbanization make for a context of change in which economic, demographic, social and cultural forces are continuously interacting with these urban spaces.

There is a third important dimension to the relationship between urban spaces and their inhabitants, and it also has a lot to do with aesthetics and identity. It is this: people 'manage' several distinct roles or identities at once. This is particularly characteristic of urban environments because of the physical and functional separation of the 'audiences' to which different roles are addressed: family, neighbours, co-workers, club members and so on. People are consequently able to present very different 'selves' in different socio-spatial contexts. The city, with its wide choice of different roles and identities becomes a 'magic theatre', an 'emporium of styles', and the anonymity afforded by the ease of slipping from one role to another clearly facilitates the emergence of unconventional behaviour. This is critical to the bohemian and neo-bohemian districts that, as we shall see in Chapter 7, are a key part of the ecology of the production of design. It is also critical to many aspects of the consumption of design. For example, as Georg Simmel noted in *The Metropolis and Mental Life*, first published in 1903, it is through fashion that individuals can ally feelings of insecurity and compensate for the overwhelming size and complexity of cities full of strangers. Fashion (and by extension other aspects of design) simultaneously allows people to communicate individuality (as they express a sense of style) and to feel a sense of security and commonality (by association with the fashion choices of others).

Summary

Design has become a central aspect of contemporary urban life; it can make things not only more attractive but also more efficient and more profitable. The fundamental importance of design to cities and urban life rests on its roles in supporting and sustaining the political economy of urbanized capitalism. Design can be seen as reflecting the *zeitgeist* of the prevailing political economy while serving, like other components of the system, as one of the means through which the necessary conditions for the continuation of the system are reproduced. Design should thus be seen as linking the economic to the cultural, as both articulating and enacting social relations and human behaviour; while designed objects, clothes, places and settings are subject to constant transformation and rewriting by both producing agents (designers, marketers and so on) and by their consumers and users.

In this context, design has been a crucial component of modernization, a product and a carrier of modernity, and a central tenet of Modernism. In the 'dream economy' of 'romantic capitalism', the design of the built environment has become intimately involved with many aspects of consumption, especially those involving an explicit design premium, such as fashion and luxury products. Design has come to alter the way people see commodities and geographic settings and has established them as 'semiotic goods' whose economic value is based in part on the meanings people give them, rather than their functionality. Design therefore facilitates the ways in which we are able to establish shared meanings and read off people's values, lifestyle and status from their possessions, the clothes they wear, and the landscapes they inhabit.

Further reading

Pierre Bourdieu (1984) *Distinction: A Social Critique of the Judgement of Taste*. London: Routledge & Kegan Paul. A seminal book in debates concerning the proliferation of consumer culture.

Alexander Cuthbert (2006) *The Form of Cities: Political Economy and Urban Design*. Oxford: Blackwell. A synthesis of recent thinking about urban design and a powerful analysis of the emergence, logic and political meaning of the built environment in different historical contexts.

Adrian Forty (2005) *Objects of Desire: Design and Society Since 1750*. New York: Thames & Hudson. A readable and well-illustrated book on design history that is grounded in social context.

Virginia Postrel (2003) *The Substance of Style: How the Rise of Aesthetic Value Is Remaking Commerce, Culture, and Consciousness*. New York: Perennial. An account of consumerism and the importance of the aesthetic attributes of places and products.

Rick Poynor (2007) *Obey the Giant: Life in an Image World*. Basel: Birkhäuser. Examines alternative ways of engaging with design, in advertising and branding.

2 Design, designers and the resurgent metropolis

This chapter examines the culture and ideology of the design professions and outlines the dynamics of the design professions within the resurgent metropolises of the contemporary, globalizing world economy. The design process is enmeshed in multilayered social relations involving multiple activities, professions and occupations. The interdependencies among design-related activities involve a broad spectrum of actors and institutions that, together, constitute 'systems of provision'. Distinctive professional ideologies are forged through designers' specialized education, professional reward systems, specialized professional literature, and the influence of the design avant-garde. These ideologies draw heavily on the history of design, with overlapping design movements creating a layered legacy of ideas. Design practitioners, meanwhile, are caught in a 'permanent contradiction' between, on the one hand, the pursuit of aesthetic ideals, and, on the other, the economic and political demands of systems of provision.

Implicated as it is in almost every aspect of contemporary urban life, playing fundamental roles in supporting and sustaining the political economy of urbanized capitalism, 'design' involves multiple activities, professions and occupations. At the broadest level, design can be thought of as the process that links the desirable with the possible. The heart of design is in conception and planning, first generating an idea and then embodying that idea in a product, whether an object, building or environment (Margolin 1998: 87). Design is about considerations of both form and function in new products, buildings, images and landscapes, drawing on technical, aesthetic and market considerations. The principal contributing disciplines and occupations include architecture, graphic design, industrial (product) design, interior design, fashion, landscape architecture, and urban design and planning. Other contributors to design include multiple aspects of engineering; event, exhibition and set design; and advertising and branding. There are complex interdependencies among all these activities. Innovation in design requires the combination of a wide range of different types of knowledge; it emerges from interactions among different actors who synthesize and recombine knowledge. Some of these activities involve higher levels of aesthetic work than others. Industrial design, for example, can typically involve significant levels of technical and scientific input. As a result, it is common to make a distinction between 'above the line' and 'below the line' design: between what is evident about an object and what is concealed from view (Dormer 1990). Styling, packaging, branding, advertising, as well as the product, building or setting itself, are above the line; market research and interpretation, models, prototypes, contract drawings, planning specifications, engineering and tooling are all below the line. Meanwhile, distinctions can also be made (Campbell 2008) in terms of 'hard' design categories (architecture, urban design, civil engineering, consumer durables, furniture, etc.) and 'soft' design categories (fashion, advertising, graphic design, magazines, etc.).

Systems of provision: structures, institutions and agents

The interdependencies among design-related activities involve a broad spectrum of actors and institutions, all embedded in time- and place-specific social relations. The specific actors vary, of course, depending on whether we are talking about architecture, fashion or industrial design; but in general will include investors, financiers, business and community leaders, and consumers as well as design professionals; while key institutions include professional associations, educational institutions, museums and galleries, government agencies and even supranational organizations. At the same time, the relations among actors and institutions need to be understood in terms of their linkages: first, within a specific local social ecology and economic structure, and second, within the broader context of economic, social and cultural change – and in particular the cultural (re)construction of the meanings of things. These sets of relations represent 'systems of provision' in the production, mediation and consumption of objects and built form (Fine and Leopold 1993; Julier 2000).

A useful theoretical framework in relation to this broader social context of design is structuration theory, developed by Anthony Giddens (1979, 1984). Structuration theory holds that human environments

> are created by knowledgeable actors (or agents) operating within a specific social context (or structure). The structure-agency relationship is mediated by a series of institutional arrangements that both enable and constrain action. Hence three 'levels of analysis' can be identified: structures, institutions, and agents. Structures include the long-term, deep-seated social practices that govern daily life, such as law and the family. Institutions represent the phenomenal

> forms of structures, including, for example, the state apparatus. And agents are those influential human actors who determine the precise, observable outcomes of any social interaction.
>
> (Dear and Wolch 1989: 6)

We are all actors, then (whether consumers or designers, members of interest groups, bureaucrats, or elected officials), and all part of a dualism in which economic, communicative, political and legal structures frame and enable our behaviour while our behaviour itself reconstitutes, and sometimes changes, these structures. Furthermore, structuration theory recognizes that we are all members of various networks of social actors: organizations, interest groups, neighbours, social classes and so on.

The way we generate meaning about objects, buildings and spaces within these networks is rooted in routinized day-to-day practices that occupy a place in our minds somewhere between the conscious and the unconscious. *Recursivity*, the continual production and reproduction of social and cultural practices through routine actions, contributes to social integration: the development of social systems and structures among agents in particular locales. Another crucial concept here is that of the *lifeworld*, the taken-for-granted pattern and context for everyday living through which people conduct their day-to-day lives without having to make it an object of conscious attention.

People's experience of everyday routines in familiar settings leads reflexively to a pool of shared meanings. Neighbours become familiar with one another's vocabulary, speech patterns, dress codes, gestures and humour, and with shared experiences of the physical environment such as streets, markets and parks. Often this carries over into people's attitudes and feelings about themselves and their locality and to the symbolism they attach to that place. When this happens, the result is a collective and self-conscious 'structure of feeling': the affective frame of reference generated among people as a result of the experiences and memories that they associate with a particular place (Heidegger 1971; Williams 1973).

Meanwhile, from Weberian sociology we are reminded that some actors and institutions can be more significant than others. Weberian analysis seeks to explain how social outcomes are influenced by institutional organization and key mediating professionals. The development of this approach in relation to contemporary processes of urbanization can be traced to the work of Ray Pahl (1969), who argued for a focus on the roles of mediating professionals such as planners, mortgage managers and realtors or estate agents. Each of these sets of key actors develops a distinctive professional ideology and value system as a result of recruitment, education and professional reward systems. Drawing on these ideologies in their day-to-day decision-making and their interpretations of social needs, priorities and market forces, these key actors can exert a decisive influence on urban out-comes, sometimes subtly, sometimes not. In some cases, their influence can be indirect, influencing people's sense of possibilities without ever becoming directly involved in an issue. The same argument can be applied to other key actors in systems of design provision: architects, industrial designers, fashion designers, and so on.

Design culture

In short, design is a process that is enmeshed in social relations. Guy Julier (2000) observes:

> No matter how much the designer tries, he or she cannot fully control the processes by which the public read, interpret, or even straightforwardly use the objects, images, and spaces they shape. . . . [T]here is a paradox here in that designers are both in command of what they do but at the same time they are the agents of ideology, subcontractors of a bigger system.

This amounts to what Julier characterizes as the 'culture of design'. His initial conception of the principal elements of the culture of design was a nodal framework in which designed objects, spaces, or images are products of – and, recursively, influences upon – the interdependencies among designers, producers and consumers (Figure 2.1).

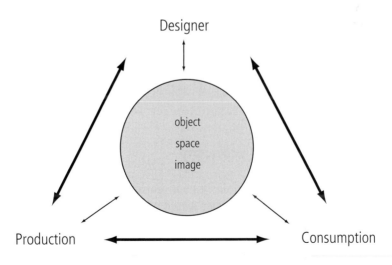

Figure 2.1 The culture of design: products of the interdependencies among designers, producers and consumers. (Source: Julier 2006)

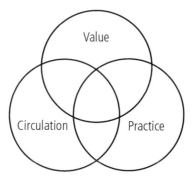

Figure 2.2 The culture of design: value, circulation and social practice. (Source: Julier 2006)

Recognizing the increasing complexity of design environments and in particular the reflexivity of both consumers and producers, Julier has moved away from this focus on designed objects, spaces and images toward the idea of overlapping subsets of design culture: value, circulation and social practice (Figure 2.2). From this perspective, we see more clearly the roles of design in the context of its broader roles in supporting and sustaining the political economy of urbanized capitalism. Julier recognizes that the designer's principal role is in the creation of value. This most obviously is commercial value, but also may include social, cultural, environmental, political and symbolic values: it is not restricted to notions of the value of 'good design'. With value established, the circulation of capital follows, mediated by knowledge networks, legislation, political pressures, economic fluctuations and fiscal policies. The overlapping arena of social practice may be conceived here in terms of Bourdieu's concept of 'fields' (see Chapter 1). Patterns of consumption, therefore, are part of practice: 'Things are bought and put to use, environments are visited, Websites are perused in fulfilling practice' (Julier 2006: 74).

Design and the cultural economy

The creation of value within design culture takes place within a broader professional and occupational milieux that is often described in terms of 'creative industries'. In the United Kingdom, the government's Department for Culture, Media and Sport (DCMS) has defined creative industries in terms of their potential to create wealth and jobs through developing intellectual property as a result of the deployment of individual creativity, skill and talent. The sectors listed by DCMS as creative industries are advertising, architecture, art and antiques markets, computer and video games, crafts, designer fashion, film and video, music, performing arts, publishing, software, and television and radio, as well as 'design' per se.

More generally, creative industries are regarded as those producing high levels of aesthetic and symbolic content relative to their functional or utilitarian properties. They are geared toward consumer demand for pleasing environments, cool and distinctive products, and for entertainment, self-affirmation and social display (Garnham 2005; Pratt 2008). As such, they are industries that drive celebrity and generate marketable products with ephemeral and elusive qualities: glamour, style, cool and trendiness. Yet they also incorporate a great deal of below-the-line expertise and rely a great deal on symbiotic relationships with a broader set of hybrid and crossover industries such as cuisine, public relations, tourism and heritage industries, hair and makeup, jewellery design and web design, along with informal creative activities such as graffiti art.

Many observers use the term 'cultural economy' in referring to the diverse and extensive array of elite and mass cultural production and distribution activities in contemporary economies, though it poses some tricky problems of definition and classification for analysts and theorists. Allen Scott (2001) provides a useful categorization within the cultural economy, describing what he calls the 'cultural products industries' as comprising

> the media (e.g. film, television, music, publishing), fashion-intensive consumer goods sectors (e.g. clothing, furniture, jewelry, and so on), many different types of services, (e.g. advertising, tourist facilities, or places of entertainment), and a wide assortment of creative professions (e.g. architecture, graphic arts, or web-page design). We may also include in these industries facilities for collective cultural consumption like museums, art galleries, or libraries, whether privately or publicly controlled.
>
> (Scott 2001: 16)

Scott suggests that, in addition to a concern with the creation of aesthetic and semiotic content, cultural products industries are generally subject to the effects of Engels' Law (meaning that as disposable income expands, consumption of the outputs of cultural-products industries rises at a disproportionately higher rate), and that they are frequently subject to competitive pressures that encourage individual firms to agglomerate together in specialized districts, while at the same time their products circulate with increasing ease on global markets (Scott 2004).

The underpinnings of the cultural economy are based on the interdisciplinarity of design. As Elizabeth Currid (2007) observes in her book on *The Warhol Economy* in New York City:

> When we think of art and culture, we often think of film and fashion, or art
> or design but often as separate entities. And while they do cultivate their own
> following, discipline and norms, they are also part of a far more encompassing
> and symbiotic whole than we generally consider them. These separate industries
> operate within a fluid economy that allows creative industries to collaborate
> with one another, review each other's products, and offer jobs that cross-fertilize
> and share skill sets, whether it is an artist who becomes a creative director for
> a fashion house or a graffiti artist who works for an advertising agency. . . .
> That the Metropolitan Museum of Art holds the Costume Institute benefit, the
> annual gala devoted to fashion design, and that Nike hires graffiti artists to
> design sneakers is evidence of the interdependent nature that artistic and cultural
> industries have with one another, and their need to be around each other and
> engaging each other in the same places.
>
> (Currid 2007: 7)

This interdisciplinarity is fostered and intensified by the social and professional
ecology that exists in certain cities. It is also fostered by the rise of branding, as
designers seek (and clients respond to) greater integration of product, graphic and
interior design in order to create coherent design solutions, and as clients seek to
diversify their product lines while exploiting endorsement by association with
signature designs and star designers. As Andy Pratt (2000) observes, this interdisciplinarity, together with long hours and socializing with co-workers and other
creative workers can result in distinctive affective communities of like-minded
workers. Such communities exhibit a distinctive habitus, their lifestyle, music,
aesthetics, decor and clothing all typically reflecting an affinity with cool, edgy
and neo-bohemian elements.

Design movements

Professionally, designers, along with many of their second cousins in the cultural
products industries, adhere to distinctive ideologies that are forged through
specialized education, professional criticism and reward systems, specialized
professional literature, the organization of firms, and the influence of the avant-
garde. These ideologies draw heavily on the history of design, with overlapping
design movements – sometimes competing, sometimes synergistic, sometimes
dialectically opposed – creating a layered legacy of ideas, assumptions, and conventional wisdoms. In Part II of the book these movements are examined in detail
with regard to their impact and influence on patterns of urban development. In
this part, a brief, broad-brush history will serve to establish the most influential
movements and indicate their legacy in terms of the ideological underpinnings of
the design professions.

The initial aesthetic response to the radical changes of the industrial revolution was reactionary. In the face of turbulence and change, designers opted for the reassurance of neoclassical styles and motifs. By the middle of the nineteenth century, designers were still struggling to find an appropriate response to industrialization and to the challenges and opportunities presented by new technologies. Neoclassicism gave way to eclectic revivals and mutations of sundry historic styles before the first authentic and meaningful design movement appeared in England in mid-century: the Arts and Crafts movement, based on a romantic idealization of pre-industrial crafts. The Arts and Crafts movement was at its height in Europe between approximately 1880 and 1910; in the United States (where it was known as the Craftsman style) it peaked between 1910 and 1925, having been occluded by another phase of reactionary aesthetics in the form of Beaux Arts and the City Beautiful movement (Figure 2.3). Important derivatives of the Arts and Crafts movement include Art Nouveau ('Jugendstil' in German-speaking countries) and Art Deco.

Meanwhile, the English Arts and Crafts movement also influenced the early pioneers of German Modernism in the 1890s and early 1900s: in the Deutsche Werkstätten, the Dresdner Werkstätten and the Debschitz School in Munich. The establishment of the Deutscher Werkbund Education Committee in 1908 laid the foundations for the Staatliches Bauhaus in 1919. The school initially promoted arts and crafts but soon made a clear and decisive shift towards industrialization and the notion of the type-form as the optimum solution to the functionality of every product.

The Bauhaus was at the heart of the broader movement of Modernism, as artists and designers grappled with the implications of a rush of technical sensations – electric power, telecommunication, internal combustion engines, aviation, radio, photography and cinematography – and the dislocations of economic change and modern warfare. Post-Impressionism, Cubism, De Stijl, Futurism, Dada, Expressionism, Constructivism, Secessionism and Surrealism all contributed to the flux of Modernism (Hughes 1980). Meanwhile, others sought to bring rationalism and science to bear on problems of design. This initially found expression in markets that were related to health and hygiene: bathroom design, for example. The most influential early advocate of a scientific approach to design, however, was Patrick Geddes, professor of biology at the University of Dundee, Scotland, between 1888 and 1919, whose analytical approach to the problems of urbanization were to become foundational to the emergent disciplines of town planning and urban design. Some strands of Modernism also gave particular expression to the rationality and scientific underpinnings of modernization. Theo van Doesburg, for example, a De Stijl protagonist, expressed his perception of a new spirit in art and design in the early 1920s:

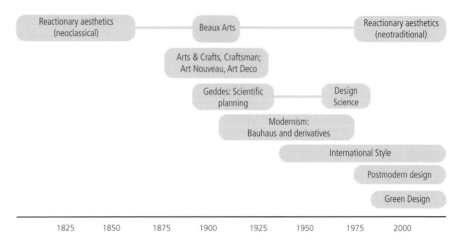

Figure 2.3 Major design movements.

> Our epoch is hostile to every subjective speculation in art, science, technology, etc. The new spirit, which already governs almost all modern life, is opposed to animal spontaneity, to nature's domination, to artistic flummery. In order to construct a new object we need a method, that is to say, an objective system.
>
> (quoted in Naylor 1968)

A little later, the architect Le Corbusier wrote about apartment buildings as rationally designed 'machines for living' and set out his ideas for *La Ville Radieuse*, a strictly rational and highly deterministic design for the ideal modern city.

Modernism, meanwhile, was to become truly international as the principal language of twentieth-century design. The decisive event was an exhibition at the Museum of Modern Art in New York City, organized in 1932 by two American architects, Philip Johnson and Henry-Russell Hitchcock. The exhibition included examples of work by former Bauhaus practitioners and by many other European Modernists. More in hope than as a reflection of reality, the exhibition was called The International Style. It did not take long, however, for the Modernism of the International Style to become accepted by corporate and institutional executives eager to install themselves in settings that were redolent of progress and prosperity. The spread of commodified Modernism in the guise of the International Style took place simultaneously with a resurgence of interest in the relevance of scientific approaches to design. In academia especially, this was driven in part by a desire to share in the legitimacy and status of the sciences, engineering and social sciences: many designers sought to shed the image of an artistic activity in favour of a scientifically based discipline. They were influenced by the likes of Buckminster

Fuller (1969), who called for a 'design science revolution' based on science, technology and rationalism to overcome the human and environmental problems that he believed could not be solved by politics and economics; and Herbert Simon (1969), who called for the development of a 'science of design' in universities, based on a body of intellectually tough, formal, analytic, knowledge about the design process.

Architects and planners sought to engage with the latest developments in social science theory and research, exploring and acquiring the languages and toolkits of behavioural theory, regional economics, regional science, quantitative geography, systems analysis and transportation modelling. Product designers held a Conference on Design Methods in London in 1962, highlighting the opportunities for using new computational methods and emerging social science theory in developing a 'scientific method' for design (Cross 2001). But the paradigm shift never occurred. As Robert Gutman (1989: 107) observed in relation to architecture, this was partly because of 'the tendency of architects to leaf through books on social science and philosophy, looking for phrases that express their personal views and lend an imprimatur to their design work'.

More important, perhaps, was that scientific approaches to design issues did not result in significant additional value to design products compared to the mystique afforded by art and the associated cult of the personality. Equally, above-the-line design problems are largely unamenable to the techniques of science, social science or engineering. They are problems that are often ill formulated, with multiple clients and decision-makers with conflicting priorities. They have been characterized as 'wicked' problems, in contrast to the precisely defined 'tame' problems of science and engineering laboratories (Buchanan 1995). Wicked problems involve the complex real-world interdependence of diverse factors and stakeholders, and therefore call for integrated and flexible design solutions based on judgement and intuition derived from routines of problem-solving that are built up through education and experience with real-world problems.

In the 1980s and 1990s a new design movement emerged from a new alliance of taste and capital. As we shall see in Chapter 5, postmodernism was antithetical to Modernism, and it developed through an alliance between representatives of the status- and consumption-oriented society of the period and the managers of newer and more flexible forms of capitalist enterprise. This was in contrast to Modernism's dependence on alliances between a cultural avant garde and corporate capital and between a liberal elite and public capital. Modernism has been reasserted to some degree with the onset of a 'second modernity', an emergent era of remodernization, this time at the global scale and involving cosmopolitanism, transnationalism and supranationalism that are at odds with the top-down managed capitalism and planned modernization of the first modernity (Beck and Lau 2005; Beck 2006). In commodified form, contemporary Modernism is often referred to as 'Euro-design', echoing its Bauhaus roots. Meanwhile, reactionary aesthetics have also been resurgent, most strikingly in neo-traditional urban design for residential subdivisions but also in revivals of regional, ethnic and period fashions in textiles, apparel and furniture. Finally, we should recognize an influential new below-the-line design movement: Green Design, characterized not by a particular aesthetic but by a concern with energy efficiency and environmental quality.

Design ideology and design practice

At heart, designers are interested in aesthetics and functionality. In practice, though, their work must be undertaken for clients who are, for the most part, driven more by profits and politics. Caught in this 'symbiotic relationship', observes Diane Ghirardo, designers tend to distance themselves from such 'hard' political and economic objectives by emphasizing their status as artists engaged in the production of aesthetically and socially meaningful form. Positioning the business of design within a 'sufficiently trivial' aesthetic frame leaves 'something innocuous at centre stage in order to divert attention from more serious concerns' (Ghirardo 1984: 114). Ghirardo was writing about architecture in particular, where the symbiotic relationship with capital is seldom addressed explicitly and is most often recast into an aestheticized 'architectural' discourse (Jones 2009).

Architectural theorist Kenneth Frampton suggests that competition among practitioners, and the constant quest for media attention, leads to 'overaestheticisation', with architects pursuing a 'succession of stylistic tropes that leave no image unconsumed, so that the entire field becomes flooded with an endless proliferation of images . . . increasingly designed for their photogenic effect' (Frampton 1991: 26). Similarly, Magali Sarfatti-Larson describes a professional ideology in which the creation of pure design is propagated as architects' purpose or raison d'être (Sarfatti-Larson 1993). But, as she notes, while professional ideology is dominated by aesthetic values, the design process is in fact highly contingent, involving not only the designers themselves but also their clients, the regulatory framework (Carmona 2009a; Imrie and Street 2009), design technologies, the design media and popular opinion. The result is that practitioners are caught in a 'permanent contradiction' between, on the one hand, the pursuit of aesthetic ideals, and, on the other hand, the 'heteronomous' social relations that make design 'an eminently political activity' (Sarfatti-Larson 1993: 14). Paul Jones (2009: 2524) points out that the romantic myth of the asocial, creative architect is particularly strong in most architectural discourse and that it serves to reinforce the conventional professional ideology with regard to, for example, aesthetic preferences (Bonta

1979), the gendered practice of architecture (Fowler and Wilson 2004), the justification of relationships with unscrupulous, powerful clients (Arnold and Hurst 2004; Sudjic 2005), the emergence of styles (Heynen 1999) and the 'silent complicity' (Dovey 1999) of many architects with regard to political and economic projects.

Design determinism

One of the most striking and recurring themes among practitioners, regardless of the design movement to which they subscribe, is a hubristic design determinism. This is especially apparent in architecture, urban design and planning, making practitioners a classic example of Weber's key actors whose day-to-day decisions are undergirded by a distinctive professional ideology. Because of the dominance of aesthetics in their professional ideology, most architects tend to see their work as abstracted from the people and places that it affects. In this sense they can be almost autistic in their determination to isolate the design process from social and political engagement (clinically, autism results in a failure to use language and perceive surroundings normally). As Dana Cuff (1989) found in her interviews with some eminent New York architects, they tend to conceive of people as beholders, not willed agents or actors. Armed with this attitude, designers have clung to an ideology that is imbued with an essentialism in which all aspects of the complex and diverse world are reduced to a set of singular and authoritative principles, summarized in a set of simple statements and strategic visual and verbal discourses. (This, of course, is a chimera: the relationships between people and their environments, and between society and material culture are in fact complex, reflexive and recursive.)

Underpinned by this determinism, much of the discourse in design circles is framed as if the author's tenets are self-evident:

> *Of course* a building must be proportioned according to principles derived from the human body (Vitruvius). *Of course* a bank must have a Renaissance façade (Beaux-Arts). . . . *Of course* we must have white walls. *Of course* there must be no decoration. *Of course* a building should express its function. Those established in the field must avoid at all costs the possibility that someone may reveal the essential arbitrariness of their aesthetic.
>
> (Stevens 1998: 99; emphases in original)

Free of the bothersome nuisances of rigorous hypothesis testing and of having to take account of the reflexivity, recursivity and messy conditionality of real-world situations, most writing about design falls within a specific formal canon, with esoteric language to lend an air of intellectual legitimacy and a self-referential logic that identifies 'good design' from bad. Coupled with the pervasive cult of the personality within design culture, the result tends to be either a cosy reaffirmation or a fratricidal quarrel that is won or lost merely on the strength of unfounded assertions. This tendency is reinforced by designers' need to differentiate themselves and become recognized as tastemakers. As a result they 'curate themselves' through the production of books, exhibitions and catalogues about their own work (Rybczynski 2008). In this subfield, appearance almost always trumps content: the artwork, typography and layout – offbeat, minimalist or funky – always crafted to create the book (or catalogue)-as-object. Designers' own writings, when they go beyond picture books with a minimum of explanatory narrative, tend to be highly stylized and often obscurantist, simultaneously oracular and opaque.

Design education

The roots of much of what is distinctive about design ideology and discourse are to be found in design education. Writing about art schools, Guy Julier (2000) observes that they tend to promote a marginalized and self-referential view of themselves:

> Once the student is enrolled, the relative lack of strict timetabling, the provision of personal studio space instead of [moving between] classrooms, the emphasis on individual creativity alongside the cultivation of a group, studio-based atmosphere conspire to produce a working practice which assumes the status of lifestyle.
>
> (Julier 2000: 36)

Design studios represent the archetype of Bourdieu's 'charismatic' mode of inculcation, whereas most of the rest of higher education follows a 'scholastic' mode, based on lectures, demonstrations, reading and writing exercises.

The leading design schools represent the discursive centres of the design disciplines and they often exercise a powerful influence on the image of their professions. Yet they tend to be somewhat distanced from the mass of practitioners. Architecture schools, in particular, tend to have a fraught relationship with the profession, resenting the restrictions on curricula imposed by professional accrediting agencies and eschewing the issues that graduates will face in the transdisciplinary and commercially oriented practices of medium- and large-sized firms (that represent 80 per cent or more of the profession) in favour of idealized notions practice in small, boutique firms committed to 'good' design.

Professional firms, for their part, tend to see the studio-based system as

> a fantasy world in which incompetent professors who are the center of petty personality cults encourage bizarrely unrealistic expectations in students, while

avoiding the teaching of anything actually to do with the hard realities of life. Students learn nothing of the other activities of the construction industry. They cannot draw and they know nothing of construction.

(Stevens 1998: 171)

Yet an architecture degree represents a liberal education as well as a professional (pre-)qualification. As Garry Stevens (1998: 3) notes, 'one of the prime functions of the system of architectural education is to produce cultivated individuals', at once a recruiting pool and a market for a wide range of cultural-products industries.

The avant-garde and hero designers

In many accounts of design history and design culture, change is accounted for by referring to the characters and careers of individual artists and designers. The role of the avant-garde is important here, though it has changed significantly over time (Bell 1976; Bürger 1984). Amid the turmoil of early modernity, avant-garde artists were able to pursue art for art's sake, pushing the limits of aesthetic experience. Although their work belonged to a 'purified' arts scene, it provided inspiration and provocation for professional designers and design educators involved in the broader project of Modernism. It did not take long before some of them became, in turn, an avant-garde of professional practice, 'hero' designers in Modernism's master narrative, in which the march of Reason in the arc of History would bring creativity, enlightenment and a progressive sensibility to a confused and reactionary world. Their manifestos and aphorisms reflected a strongly idealistic, utopian and deterministic attitude, while their individual careers thrived on a radical egocentricity that was often carefully cultivated to enhance their 'heroic' image. This, in turn, went a long way toward cementing the cult of the personality as a key dimension of design ideology.

The most notorious example is Le Corbusier, whose work is discussed in Chapter 4. Born in 1887 as Charles-Edouard Jeanneret-Gris, he changed his name to Le Corbusier in a meaningless but brilliant stroke of personal branding. A determined self-publicist, even his signature was designed, while his distinctive spectacles were to become a global shorthand for 'architect'. Undoubtedly a talented artist and architect, his professional success and notoriety were also a result of his mono-maniacal, narcissistic and pugilistic temperament and his deeply authoritarian, opportunistic and cynical approach to professional work (Weber 2008). Unfortunately, his behaviour as well as his ideas became something of a role model for lesser talents. Meanwhile, other protagonists of the artistic avant-garde were concerned more with challenging the status quo in the cause of social, political, and economic reform. Dada and Surrealism, in particular, sought to use art to transform life: art should not be limited to its own small sphere, it should revolutionize society. The ultimate aim of Dada and Surrealism was to integrate art into society and everyday life as the enemy of the instrumental rationality that serves the political economy of capitalism. This tradition of the avant-garde still exists, but only in a marginalized and somewhat romanticized way. It has been characterized by Peter Bürger (1984) as the 'historical avant-garde', superseded by neo-avant-gardes who, in Bürger's view, have merely recycled the forms and strategies devised by their predecessors, reaping huge institutional and commercial success without any real struggle for change. What lingers, therefore, is the figure of the 'hero designer' that circulates through educational curricula and the popular media. As Suzanne Reimer and Deborah Leslie note, the hero designer figure today is in part a relatively simplistic media construction. 'For readers or viewers not necessarily well versed in design ideas and terminology, stories about individuals enable journalists to use the character of a designer to convey a popular story' (Reimer and Leslie 2008: 150). The expansion of design publishing has helped to construct increasingly knowledgeable and reflexive consumers who have become interested in 'design classics'. This has created an expanded market for (relatively) inexpensive copies of the work of established design heroes like Le Corbusier, Charles Eames, Arne Jacobsen, George Nelson and Mies van der Rohe and an appetite for the work of up-and-coming young designers who are shaping up as the new heroes – not of social transformation, but of the marketplace. Leading designers, increasingly, are designing themselves, creating distinctive identities that are amenable to corporate branding and the global consumer economy. The exemplar here is Rem Koolhaas (see Case Study 2.1).

Globalization, the new economy and the resurgent metropolis

The global reach of practitioners like Rem Koolhaas is symptomatic of what Robert Reich (2007) calls 'supercapitalism', a contemporary political economy dominated by transnational corporations at the expense of public institutions, community values and local cultures. To some observers, the shifts involved in the rise of global capitalism – supercapitalism, in Reich's terminology – amount to an epochal change: what Ulrich Beck (Beck and Lau 2005; Beck 2006) has argued as the onset of a second modernity. Whereas economic and urban development during the first modernity was framed by competitiveness within closed geographic systems (national states) that were, in turn, competing with one another, urban development at the beginning of the second modernity is subject to competitiveness at the global scale. Meanwhile, in attempts to recapture some control over the scale of the new economic logic and its social, cultural and environmental implications, national governments have become increasingly collaborative, supranational entities have emerged, many institutions have extended their focus from a national to an international frame of reference, and many local and regional organizations have become involved in cross-border collaborative networks of one sort or another. An increasing public awareness of the complex, multiscalar interdependence that characterizes the second modernity (underscored by the consequences of global warming, by energy costs, by food scares and food shortages, and by the 2008–2009 global financial 'meltdown') has contributed to new sensibilities. Beck himself (1992, 2008) has emphasized the emergence of a 'risk society' preoccupied with concerns about security and sustainability.

For more and more people in the design professions, globalization has radically reconfigured the nature of work itself. Andrew Jones (2008) writes of 'global work', pointing out that working practices, the nature of workplaces, the experience of work, and the power relations in people's working lives are increasingly influenced by the dynamics of the global economy. Leslie Sklair (2005) writes of the 'transnational capitalist class', people who operate transnationally as a normal part of their working lives and who more often than not have more than one place that they can call home. There are four distinct fractions of this class, according to Sklair, each involving the principals and employees of different kinds of design firms. Sklair's particular focus is on architecture:

1   *The corporate fraction*: the major transnational corporations and their local affiliates. In architecture these are the major architectural and architecture–engineering firms, the likes of URS Corporation, Nikken Sekkei, Ellerbe Becket, Gensler, and Skidmore, Owings & Merrill (SOM).
2   *The state fraction*: globalizing politicians and bureaucrats at all levels of administrative power and responsibility who actually decide what gets built where, and how changes to the built environment are regulated. This fraction is increasingly important as cities compete for global status through promotion of iconic architecture.
3   *The technical fraction*: globalizing professionals, ranging from the leading technicians centrally involved in the structural features and services (including financial services) of new buildings to those responsible for the education of students and the public in architecture. Examples of major firms with employees in this fraction include Arup (a global engineering, design, planning and business consulting firm) and Cushman and Wakefield (a global real estate research, investment and consulting firm).
4   *The consumerist fraction*: retailers and media responsible for the marketing and consumption of architecture. Especially notable here is the connection between architecture and shopping, as illustrated by Prada's commissioning of signature architects like Koolhaas, Herzog & de Meuron, and Kazua Sejima to design iconic stores for them in globalizing cities, and by consumer magazines like *Wallpaper**.

Case Study 2.1   **Rem Koolhaas: branded design hero**

Rem Koolhaas began his practice in London in 1975 with partners Elia and Zoe Zenghelis and Madelon Vriesendorp. In contrast to the usual convention of naming a practice after the principals, they called their firm the Office for Metropolitan Architecture (OMA), presaging the inspired branding that has characterized the firm's activities ever since. OMA took a long time to become established, but fame came quickly to Koolhaas himself as a result of his book, *Delirious New York* (Monacelli Press, 1978). Koolhaas had spent a year in New York and his book was in some ways a classic of architectural determinism: New York's vitality and metropolitan lifestyle are part of a 'culture of congestion' resulting from the city's high-density grid layout. But what was new was the style: breathless, iconoclastic and atheoretical, filled with fascinating (but often disconnected) facts and striking and unusual images, and studded with catchy new labels ('architectural mutations', 'utopian fragments') for the established landmarks of New York's built environment.

After a series of relatively small commissions in the Netherlands, OMA moved to Rotterdam, and the firm's breakthrough came in 1994 with a bold master plan for Lille, in northern France, designed to take advantage of the city's strategic location within the emerging European high-speed rail network. Since the mid-1990s Koolhaas has become increasingly sought-after as a 'signature' architect, winning important commissions and producing a series of award-winning buildings, including the Educatorium on the campus of Utrecht University (1997), the Maison à Bordeaux (1998), the Netherlands Embassy in Berlin (2003), Seattle Central Library (2004), Seoul National University Museum of Art (2005), Shenzhen Stock Exchange (2006) and the Central China Television building in Beijing (2009 – Figure 2.4). He was awarded the Pritzker Prize for architecture (the profession's most prestigious international award for practice) in 2000.

This success coincided with an unprecedented amount of branding and self-promotion. Much of it was facilitated through the creation in 1999 of a sister company, AMO (not an acronym for anything, simply a cute play on the acronym of the parent firm). AMO is described as a 'think tank' and research studio for design, allowing Koolhaas to claim intellectual territory and expertise in areas well beyond the boundaries of architecture. Through AMO, Koolhaas has framed his image as pragmatic and purposive, engaging directly with the spectacles and realities of contemporary urbanization and consumer culture.

In parallel with the firm's branding, Koolhaas has produced a series of publications that have raised his own profile and contributed to his image as both a designer

Figure 2.4 The Central China Television building in Beijing, designed by Rem Koolhaas. (Photo: Tom Fox/Dallas Morning News/Corbis)

and insightful cultural intermediary. *S, M, L, XL* (with Hans Werlemann and Bruce Mau; Monacelli Press, 1994) was a series of stories, declarations and justifications of the first decade of OMA's work. His contribution to *Mutations* (with Stefano Boeri, Sanford Kwinter, Daniela Fabricius, Hans Ulrich Obrist and Nadia Tazi; Arc en rêve centre d'architecture, 2001) was a result of his engagement with the Harvard Project on the City, aimed at explicating the visual outcomes of contemporary urbanization. Koolhaas took a teaching position at Harvard on the condition of not having to teach design, so he could focus on broad topics like shopping and on the contemporary shock-cities of Lagos and the Pearl River Delta. These projects led to *The Great Leap Forward* (with Bernard Chang, Mihai Craciun, Nancy Lin, Yuyang Liu, Katherine Orff and Stephanie Smith; Taschen, 2002) and *The Harvard Design School Guide to Shopping* (with Chuihua Judy Chung, Jeffrey Inaba and Sze Tsung Leong; Taschen, 2002). These books all built on the successful affect of *Delirious New York*, with garish, splashy graphics, grainy photoreportage, an assortment of demographic and economic statistics, and essays that pivot around catchy labels ('Generica', 'Junkspace') and would-be profundities ('World Equals City'). Meanwhile, and in similar vein, OMA-AMO published *Content* (Taschen, 2003), part book, part magazine, part 544-page Koolhaas fanzine. In 2003,

Koolhaas was Guest Editor of the June edition of *Wired* magazine, introducing his expansive ideas to its techno-utopian, new-economy readership. In 2005, Koolhaas co-founded *Volume* magazine, which claims to set the agenda for design. As the magazine's own website puts it with blithe pomposity:

> going beyond architecture's definition of 'making buildings' [the magazine] reaches out for global views on designing environments, advocates broader attitudes to social structures, and reclaims the cultural and political significance of architecture. Created as a global idea platform to voice architecture any way, anywhere, anytime, it represents the expansion of architectural territories and the new mandate for design. . . . an experimental think tank devoted to the process of real-time spacial [*sic*] and cultural reflexivity.

As both a designer and cultural intermediary, Koolhaas has surfed (rather than challenged) contemporary economic and cultural trends. His forays into the 'dirty realism' of 'Generica', 'Junkspace', and the wild and unrestrained urbanism of Lagos and the Pearl River Delta are deployed as counterpoint to the conventional professional canon; but this serves only to mask the structural forces that underpin the visual and social outcomes of global capitalism. His design work, on the other hand, unequivocally embraces the dominant economic and cultural flows of the global consumer economy. His success in branding himself as a cultural intermediary, meanwhile, has elevated Koolhaas from design-hero to public intellectual. In October 2008 he was invited to become a member of a European 'Council of the Wise' under the chairmanship of former Spanish Prime Minister Felipe Gonzalez to help 'design' the future European Union in relation to long-term challenges such as climate change, globalization, international security, migration, modernizing the European economy and strengthening the European Union's competitiveness.

Key readings

Donald McNeill (2009) 'Rem Koolhaas and Global Capitalism', in McNeill, D., *The Global Architect: Firms, Fame and Urban Form*. London: Routledge.

William Saunders (2008) 'Rem Koolhaas's Writing on Cities: Poetic Perception and Gnomic Fantasy', in Kelbaugh, D. and McCullough, K. (eds) *Writing Urbanism: A Design Reader*. London: Routledge.

Michael Sorkin (2005) 'Brand Aid; or, The Lexus and the Guggenheim', in Saunders, W.S. (ed.) *Commodification and Spectacle in Architecture*. Minneapolis, MN: University of Minnesota Press.

The transnational capitalist class is part of a 'new economy' based on technology-intensive manufacturing, services (business, financial and personal), cultural-products industries (such as media, film, music and tourism) and design and fashion-oriented forms of production such as clothing, furniture, product design, interior design and architecture. As Allen Scott (2001) suggests, these and allied sectors have now largely supplanted mass-production industries as the main foci of growth and innovation in the leading centres of world capitalism. The new economy is increasingly dominated by large transnational corporations and intimately tied in to complex flows of information and networks of commodity processing, manufacture and sales that are global in scope.

A distinctive feature of the new economy is that higher-income earners have emerged in occupations that have previously had only a weakly established social status. A 'new bourgeoisie' has emerged, consisting of 'symbolic analysts': economists, financial analysts, management consultants, personnel experts, marketing experts, purchasers and, of course, designers. They have been joined by a 'new petit bourgeoisie' dominated by well-paid junior commercial executives, engineers, skilled high-technicians, medical and social service personnel, and people directly involved in cultural production: authors, editors, radio and TV producers and presenters, magazine journalists and the like (Bourdieu 1984).

Scott Lash, Ulrich Beck and others refer to these classes as the 'advanced services middle classes' – the principal patrons of symbolic consumption and the innovative class fraction associated with the 'reflexive modernization' that is characteristic of the second modernity (Beck *et al.* 1994; Lash and Urry 1994). Soja (2000) refers to them simply as 'Upper Professionals', noting that this group 'demands much more and has the public and private power to make its demands fit into the crowded, edgy, and fragmented built environment, increasingly shaping the city building process to their own image' (Soja 2000: 276).

The point here is the significance of the new economy in terms of patterns of consumption as well as patterns of production. The affluent new class fractions of the new economy have become a research and development lab for consumer preferences as well as the promoters of an intensified and voracious consumption ethic. Houses, neighbourhoods, interior design, clothes, gadgets, food – everything – is now freighted with meaning, a consequence of the aestheticization of everyday life described in Chapter 1. In addition, the space-time compression of the new economy and the associated globalization and homogenization of popular culture has fostered the perceived need for distinctiveness and identity: 'this society which eliminates geographical distance reproduces distance internally as spectacular separation' (Debord 1990: para. 167). In Debord's 'society of the spectacle', where the emphasis is on appearances, the symbolic properties of urban settings and material possessions have come to assume unprecedented importance.

Cities as crucibles of creativity and design

Allen Scott (2008) points to another significant consequence of the growth and development of the new economy (or, as he suggests calling it, the new cognitive-cultural economy): the resurgence of a distinctive group of metropolitan areas that are now forging ahead on the basis of their command of the new economy, their ability to exploit globalization to their own advantage, and the selective revital-ization of their internal fabric of land use and built form. In these metropolises

> distinctive clusters of firms in this new economy congregate together in specialized industrial districts within the fabric of urban space where they typically also exist cheek by jowl in association with a range of allied service suppliers and dependent subcontractors.
>
> (Scott 2008: 554)

These specialized districts act as 'creative fields', distinctive settings rich with innovative energies, dense interpersonal contacts and informal information exchanges. As Scott observes, much of the information that circulates in this manner is little more than random noise:

> Some of it, however, is occasionally of direct use, and discrete bits of it – both tacit and explicit – sometimes combine together in ways that provoke new insights and sensibilities about production processes, product design, markets and so on.
>
> (Scott 2008: 555)

In this way, the creative-field effects within metropolises with concentrations of new-economy industries facilitate individually small-scale but cumulatively significant processes of learning and innovation, underpinning their resurgence within the global economy. Cities that have been most caught up in these processes not only have experienced profound changes in their economic and demographic profiles, but also have undergone dramatic transformations in their physical appearance. These include gentrification, branded neighbourhoods, large-scale urban regeneration projects, iconic buildings and 'semiotic districts' (Koskinen 2005), specializing in goods and services with high semiotic content: flagship stores, megastores, shops-in-shops, high-end restaurants, cafés, art galleries, antique stores and luxury retail shops.

As David Harvey points out (1998: xi), large cities have in fact long been recog-nized as 'important arenas of cultural production, forcing-houses of cultural innovation, centres of fashion and the creation of "taste"'. Howard Becker (1984) has described how the 'art worlds' in major cities depend on a dense ecology of fellow artists, buyers and patrons, dealers, art schools, critics, its bars and hang-outs, galleries and museums, gallery owners and collectors, affordable studio spaces, equipment suppliers and specialized technicians. Others have stressed the importance of the 'creative buzz' in the social production of knowledge and diffusion of innovation in design and design-related professions (Currid 2007).

As Jane Jacobs emphasized in her famous book on *The Death and Life of Great American Cities* (1961), density and diversity generate serendipity, unexpected encounters and 'new combinations' that sometimes lead to innovation. Peter Hall, in *Cities in Civilization* (1998), wrote about the importance of a creative milieu to the 'golden ages' of Athens, Rome, Florence, Vienna, London, Paris, New York and Los Angeles, among others. Creative milieux seem to have some things in common, including a certain density of communication, which seems to require a rich, old-fashioned, even overcrowded, traditional kind of city. Creative milieux are quintessentially chaotic but culturally many-sided: rich in fundamental knowledge and competence, with good communications both internally, through close physical proximity, and externally. The synergy comes from variation and diversity among activities that are often small scale (Hall 2000).

The agglomeration inherent to cultural milieux is also important because it can advance a city's identity as a centre of design. As Currid (2007: 157) puts it: 'It is not just that the product brands the place and the place brands the product. Just as important, particular places actually dictate global taste.' Certain cities derive a kind of monopoly rent as a result of the image they acquire from the particular products and firms they are associated with: fashion and graphic design in New York; architecture, fashion and publishing in London; furniture, industrial design and fashion in Milan; haute couture in Paris; sportswear and athletic shoes in Portland, Oregon, and so on. Favourable images create entry barriers for products from competing places. Cultural agglomerations in the likes of London, Paris, Milan, New York and Los Angeles establish the cities as global tastemakers – design objects in themselves – and cultural producers want their products associated with them. Meanwhile, each cultural milieu, even those in much smaller cities, tends to have a distinctive atmosphere and buzz derived from the particular national and regional culture, the particular mix of cultural and design-related activities within it, and the local taste systems of the gentrifiers and 'cultural tribes' attracted to the area (Drake 2003; Rantisi 2004; Reimer *et al.* 2008).

The importance of local design cultures and the attractiveness of urban cultural milieux to symbolic analysts/advanced services middle classes/upper professionals has not escaped city governments, many of which have pursued policies aimed at facilitating gentrification and adopted intensive city branding campaigns that draw on their image as centres of design and creativity and on the imagery of their creative districts and iconic buildings. These are topics that are explored in more detail in Chapter 5.

Meanwhile, we should note the major policy fad in the United States that was prompted by Richard Florida's (2002, 2005) work on what he called the 'creative class' – a broadly defined group of new-economy symbolic analysts/advanced services middle classes/upper professionals. Noting that a significant positive correlation exists between the incidence of the creative class in different cities and local economic growth, Florida's argument is that urban economic development will depend increasingly on cities' ability to attract and retain significant numbers of this mobile but choosy class. The inference, he suggests, is that urban policy should focus on providing the right 'people climate' in order to attract mobile creative talent. This means, he argues, promoting social and ethnic diversity, investing in high-quality cultural amenities and urban design, and cultivating bohemian and 'edgy' urban environments conducive to modish lifestyles.

It is an argument that has proved to be very attractive to civic leaders across the United States and beyond, though as Allen Scott (2006: 11) observes, Florida 'fails signally to articulate the necessary and sufficient conditions under which skilled, qualified, and creative individuals will actually congregate together in particular places and remain there over any reasonably long-run period of time'. Jamie Peck (2005: 763) notes that 'rather than "civilizing" urban economic development by "bringing in culture", creativity strategies do the opposite: they commodify the arts and cultural resources, even social tolerance itself, suturing them as putative economic assets to evolving regimes of urban competition'.

Summary

Design is about considerations of both form and function in new products, buildings, images, and landscapes. Although professional ideology and discourse is dominated by aesthetics, the design process is in fact highly contingent, involving not only the designers themselves but also their clients, the regulatory framework, design technologies, the design media and popular opinion. The interdependencies among design-related activities involve a broad spectrum of actors and institutions, all embedded in time- and place-specific social relations. Structuration theory provides a useful theoretical framework in relation to this broader social context of design.

The 'systems of provision' in the production, mediation and consumption of objects and built form commonly involves a great deal of interdisciplinarity, and this in turn is fostered and intensified by the social and professional ecology that exists in certain cities. It is also fostered by the increasing importance of branding, as designers seek (and clients respond to) greater interprofessional integration in order to create coherent design solutions, and as clients seek to diversify their product lines while exploiting endorsement by association with signature designs and star designers.

Nevertheless, many designers distance themselves from 'hard' political and economic issues by emphasizing their status as artists who are engaged in the production of aesthetically and socially meaningful form. This stance is often associated with a strong adherence to a somewhat simplistic and deterministic perspective, something that is in turn influenced by the cult of personality and the artistic avant-garde. Meanwhile, for more and more people in the design professions, globalization has radically reconfigured the nature of work. Leading design professionals are part of a 'transnational capitalist class' that has been produced by the growth of the new economy. This new class fraction tends to be concentrated in a distinctive group of metropolitan areas that are now forging ahead on the basis of their ability to exploit globalization to their own advantage. In these metropolises, distinctive clusters of firms congregate together in specialized districts that act as 'creative fields', distinctive settings rich with innovative energies, dense interpersonal contacts and informal information exchanges.

Further reading

Peter Hall (1988) *Cities in Civilization*. London: Weidenfeld & Nicolson. A wide-ranging survey of the roles of markets, technology, culture and politics in the creativity and innovation associated with the 'golden ages' of major cities.

Guy Julier (2000) *The Culture of Design*. London: Sage. An important book on the design process and how design disciplines act and interact in the world. It has a good balance between theoretical material and detailed analysis of illustrative examples.

Allen J. Scott (2008) 'Resurgent Metropolis: Economy, Society and Urbanization in an Interconnected World', *International Journal of Urban and Regional Research*, 32.3: 548–564. A seminal article on the emergence of a new kind of urban economic dynamic involving both globalization and the growth of cognitive-cultural production systems.

Leslie Sklair (2005) 'The Transnational Capitalist Class and Contemporary Architecture in Globalizing Cities', *International Journal of Urban and Regional Research*, 29.3: 485–500. Describes the role of each of the four fractions of the transnational capitalist class in the production and consumption of architecture in world cities.

Garry Stevens (2002) *The Favored Circle: The Social Foundations of Architectural Distinction*. Cambridge, MA: MIT Press. Based on a theoretical framework that draws heavily on the work of Pierre Bourdieu, this book examines the social and professional context of architectural education and practice.

PART II
The intentional city

In Part II, the focus is on the imprint of design on cities: the 'Intentional City' of grand designs, paternalistic reforms, utopian ideals, new towns, modernization programmes, monumental projects, city plans, renewal and regeneration schemes, imagineered environments, and private master-planned communities – urban settings that have been created with a degree of intentionality, whether by land-owners, developers, governments, planners, urban designers or some combination. The three chapters in this part of the book are arranged in broad chronological sequence, beginning with key historical antecedents and following through to contemporary attempts to introduce principles of sustainability into urban design. The focus is on the period since the industrial revolution. The intention is not to be comprehensive – a multivolume task – but, rather, to show the evolving, inter-dependent relationship between processes of urban change and approaches to urban design and planning, and their consequences, intended and unintended.

3 **Better by design?**

Historical antecedents

This chapter provides a brief introduction to the imprint of design and design-related policies on cities, tracing the interactions between changing urban conditions and key interventions in architecture, urban design and planning. Early antecedents are traced to the Renaissance and Baroque periods, when architecture and urban design gave expression to reason, rationality and idealism. The principal focus of the chapter, however, is the industrial era and the entirely new set of issues that emerged to both stimulate and challenge design. For the affluent middle classes, the industrial revolution brought new levels of material consumption, and new settings in which to purchase and consume them. The situation of the poor, meanwhile, was gradually addressed by a series of reform movements and philanthropic initiatives in a struggle to establish order, safety and efficiency. Finally, the new technologies and socio-cultural context of the industrial era gave rise to new sensibilities that led to the emergence of modern design.

The form of cities has been influenced by design since the earliest times, though the motivation has varied a great deal, from mythology and religion to geopolitics, military strategy, national identity, egalitarianism, public health, economic efficiency, profitability and sustainability. Similarly, the driving forces behind urban design and planning have ranged from despotic powers to utopian idealists, and from democratic governments to private developers. The earliest antecedents can be traced to the centralized power of early empires. Roman town planning was characterized by rectilinear street layouts that were oriented to the cardinal points of the compass. The latter was a result of Roman mythology, while the grid layout was a practical solution to laying out new settlements as the empire expanded geographically. Greek city-states were designed with rectilinear street patterns for the same reason. Meanwhile, both Greek and Roman cities shared another aspect of design that derived from centralized power: imposing symbolic and institutional structures, systematically arranged in the urban core – the *forum* of the Roman *urbs* and the *agora* of the Greek *polis*.

The Medieval period was dominated by slowly changing, introverted feudal systems, with little innovation in terms of urban design and planning. Beyond ecclesiastical grounds and their gothic architecture, there was little by way of intentional urban development apart from defensive walls and the palaces and castles of ruling families: so that medieval towns and cities were characterized by narrow, winding, 'organic' street patterns with a naturally evolved neighbourhood functional differentiation according to particular trades and their guilds. The exceptions were the new towns of politico-military origins, established, as Henri Pirenne (1952: 81) put it, as Europe 'colonized itself' in the twelfth and thirteenth centuries, when local lords opened up frontier regions, cut down forests and drained marshlands. Typically, the central authority paid for and organized the town's defences and controlled the layout of the town, usually some variant of the grid system. People were induced to settle in these new towns by the grant of a house plot within the town together with farming land in the vicinity or other economic privileges. In return, the central authority was able to pacify and control the surrounding region from the town's military garrison, and to generate revenue from taxes and market fees imposed on residents and traders.

Among the best known of these planned or 'planted' towns are the French 'bastide' towns of Aigues-Mortes and Carcassonne. In addition, there were 135 or so other planted towns in Europe in this period, including Castelfranco di Sopra, Scarperia, and Terranuova in the Florentine Republic; Offenburg and eleven other towns along the Rhine between what are now Switzerland and Germany; and a similar number of towns in England and Wales including Caernarvon, Flint, Kingston upon Hull, Liverpool, Ludlow, Portsmouth, Salisbury and Winchelsea (Morris 1994). Ludlow, for example, came into existence as a planted town after the Norman Conquest of 1066. It was established as one of a series of outposts guarding the Welsh Marches, or borderlands. Walter Lacy I, one of a group of knights who followed William Duke of Normandy across the English Channel, was given an unusually large number of manors (over 200) in the Marches in return for guarding the territory against the Welsh.

Between 1086 and 1094 Walter Lacy and his son Roger built a castle above the steep slopes of the Teme Gorge, founding Ludlow as a strategic garrison town. But while the motives of the Lacys in planting a town at Ludlow may initially have been strategic, they also saw the town as an investment that could yield income. A planned town was laid out around the castle, partly to provide essential services for the garrison, partly to stabilize the surrounding countryside (still seething with anti-Norman sentiments) and partly as a source of income for the Lacy family through market tolls, rents and court fines. In this regard, Ludlow was part of a deliberate and fairly widespread policy of town plantation by the Normans, a policy that coincided with the unprecedented boom in urban growth in the twelfth and thirteenth centuries as increased trade across northern Europe marked the transition from feudalism to merchant capitalism. As a planned town, Ludlow's streets were laid out to spacious dimensions in a basic grid pattern, with a secondary system of parallel streets and service lanes and an exceptionally wide marketplace – strikingly different from the narrow, winding lanes of most medieval towns.

Renaissance and Baroque ideals: grand design

The roots of modern Western urban planning and design can be traced to the Renaissance and Baroque periods in Europe, when artists and intellectuals dreamed of ideal cities, and rich and powerful regimes used urban design to produce extravagant symbolizations of wealth, power and destiny. Inspired by the classical art forms of ancient Greece and Rome, Renaissance urban design sought to recast cities in a deliberate attempt to show off the power and the glory of the state and the Church. The competition-winning designs of Lorenzo Ghiberti for new bronze baptistery doors at Florence Cathedral in 1401 are generally taken as the first expression of the Renaissance in the arts. The first Renaissance architecture is similarly seen to be the Foundling Hospital designed by Filippo Brunelleschi in 1419, also in Florence. The earliest Renaissance urbanism – the conscious arrangement of buildings into a predetermined, scenographic form – is considered to be the Via Nuova in Genoa, that dated from 1470.

The development of the arts in the Renaissance was closely linked with the subsequent emergence of literary and scientific humanism. The arts established an intellectual context favourable to the eventual eclipse of reactionary medieval mysticism. The field of architecture and urban design gave expression to reason, rationality and idealism, and the theories and ideas of leading architects and artists such as Leon Battista Alberti (1404–1472), Francesco di Giorgio Martini (1429–1502), Andrea Palladio (1508–1580), Pietro Cataneo (1510–1569) and Leonardo da Vinci (1452–1519), along with the rediscovered writings of Roman scholar Vitruvius, quickly spread through the new medium of printing. As Morris (1994) observes:

> By the time that Renaissance attitudes and style had been firmly established, the new technique of printing enabled new designs and theories to be communicated internationally; it was no longer necessary to turn ideas into buildings to demonstrate architectural intentions and to influence others. . . . From the fifteenth century onwards there was a succession of published works dealing with the theory of architecture, urban design, and military engineering.
>
> (Morris 1994: 165)

Spreading slowly from its origins in Italy at the beginning of the fifteenth century, Renaissance design had diffused to most of the larger cities of Europe by the end of the eighteenth century. Naturally, ideas about architecture and urban design changed significantly over these three centuries. Architectural historians typically divide the period into four phases: Early Renaissance (1420–1500), Late Renaissance (1500–1600), Baroque (1600–1750) and Rococo or neoclassical (1750–1900). The entire period is characterized by discipline and order in architecture, with generally applied rules of proportion governing the organization of urban space and the three-dimensional massing and detailed facades of buildings. Figure 3.1 shows a chronology of examples of key buildings and urban design elements in Europe.

In contrast to the Gothic designs of the Medieval period, emphasis was placed on the horizontal plane rather than the vertical. Renaissance art and design is generally associated with attempts to express calm and beauty, balance and regularity. Baroque art and design, on the other hand, 'wants to carry us away with the force of its impact, immediate and overwhelming' (Wölfflin 1966). Baroque urbanism strived for a sense of infinite space and for imposing grandeur, and was only possible on any scale where centralized, autocratic power combined with sufficient economic resources to instigate and implement complex planning programmes on hitherto unheard-of scales, most notably those of Louis XIV and Louis XV at Versailles, Peter the Great at St Petersburg, Pope Sixtus V in Rome and, at the very end of the period, L'Enfant's plan for Washington, DC. Rococo design emphasized luxury and opulence, using rich, complex and intricate decoration of interiors and exteriors, characterized by curvilinear and asymmetric patterns and exotic motifs of shells, dragons, palm trees and plants.

Dramatic advances in military ordnance (cannon and artillery) in the Early Renaissance brought a surge of planned urban redevelopment that featured impressive fortifications: geometric-shaped redoubts, or strongholds and an extensive *glacis militaire* – a sloping, clear zone of fire. Inside new walls, cities were recast according to a new aesthetic of grand design – fancy palaces, and geometrical plans, streetscapes and gardens that emphasized views of dramatic perspectives. These developments were often of such a scale that they effectively fixed the layout of cities well into the eighteenth and even into the nineteenth centuries, when walls and/or glacis eventually made way for urban redevelopment in the form of parks, railway lines, beltways or ring roads.

Meanwhile, the Renaissance coincided with marked increases in the population of European cities. As a result, hemmed in by defensive fortifications, there were few opportunities for comprehensive urban redevelopment. Even when cities were extensively damaged by fire – as happened with the Fire of London in 1666 – there

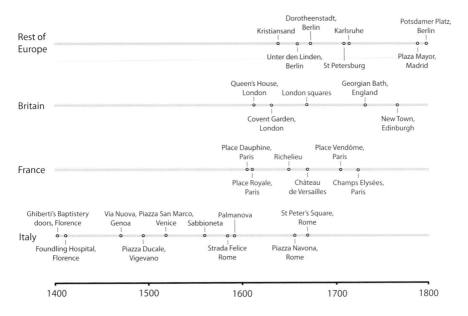

Figure 3.1 Early innovations and interventions in Europe.

was neither the political will nor the bureaucratic planning apparatus to impose extensive new plans. The relatively few planned settlements of the fifteenth to eighteenth centuries were therefore primarily either of a strategic military origin (e.g. Palmanova and Sabbioneta in Italy, Neuf Brisach in France and Kristiansand in Norway); or the result of autocratic rule, as in Richelieu and Versailles in France, and Karlsruhe in Germany.

St Petersburg, the only example of a major city founded during the Renaissance period, was a result of both military strategy and the autocratic power of Peter the Great (1682–1725), who founded St Petersburg and developed it as the planned capital of Russia on a swampy site at the mouth of the River Neva. Like Venice, St Petersburg rests on countless wooden piles to prevent it from subsiding into its marshes. It was built at the cost of thousands of human lives, but the tsar was determined to create an imperial capital to rival those of continental Europe. He also wanted St Petersburg to be Russia's 'window on Europe', exposing Russia to the new ideas and technology of the Renaissance. Often called the 'Venice of the North' because of the opulence of its architecture and its canals, St Petersburg was deliberately fashioned in the Grand Manner as a Western European-style capital city. The tsars' architects were able to lay out their work unrestricted by any legacy of old streets or buildings. Over two centuries, they collectively created a mar-vellous set piece of urban design, with imposing public buildings, imperial palaces and churches in the Baroque, Rococo and classical styles, all laid out around impressive plazas and along broad boulevards. Today, the city's imperial past is still very visible in the Grand Design of the core area on the south bank of the River Neva – the Palace Square, the Admiralty building with its landmark elegant spire and the Winter Palace, which now houses the Hermitage Museum, a treasure house of fine art of worldwide significance that originated in 1764 as the private collection of Catherine the Great.

Elsewhere, Renaissance urban design was restricted to the regeneration of major public buildings, the restructuring of primary city streets, the creation of enclosed squares, plazas and piazzas, and the addition of extensive new residential districts, typically laid out to a rectilinear plan. Important exemplars of restructured primary streets include Unter den Linden in Berlin; the Via Leonina (now the Via di Ripetta) and the Strada Felice in Rome; George Street and Princes Street in Edinburgh; and the Champs Elysées in Paris. Important exemplars of enclosed squares, plazas and piazzas include the Place Dauphine, the Place Vendôme and the Place Royale (renamed the Place des Vosges after the French Revolution) in Paris; Potsdamer Platz and Leipziger Platz, in Berlin; Covent Garden, Leicester Square (Figure 3.2) and Bloomsbury Square in London; the Piazza Navona in Rome; the Plaza

Figure 3.2 Leicester Square, London, c. 1750. Initially developed in the 1670s by Robert Sidney, second Earl of Leicester, an early example of the format of private property development in central London. (Credit: Corbis Historical Picture Library)

Figure 3.3 Piazza Ducale, Vigevano, Italy. The plan of the piazza is based on regular geometric forms and the facades of the surrounding buildings are based on a rhythm of repeating elements. (Photo: author)

Mayor in Madrid; the Piazza San Marco in Venice; the Piazza Ducale in Vigevano, Italy (Figure 3.3). Important exemplars of new residential districts include Dorotheenstadt and Friedrichstadt in Berlin, the New Town in Edinburgh (Case Study 3.1) and Georgian Bath.

Case Study 3.1 **Edinburgh's New Town**

Unlike most of Europe's cities, Edinburgh has no Roman origins. The first settlement, Dunedin, was Celtic, and took advantage of a defensive site on higher ground above the southern shores of the Firth of Forth. By the sixteenth century, Edinburgh had become the capital of Scotland, adding court and ecclesiastical functions to those of its mercantile economy. The lineaments of its development had also been established, with the main thoroughfare – the High Street – running eastwards from the castle site on a volcanic crag and a population of around 30,000 crammed along the adjacent ridge. Although the city grew steadily in population over the next century or so, development was largely restricted to the High Street ridge because of the surrounding marshy ground on both sides and the precipitous face of the sills of Salisbury Crags at its eastern end. By the early eighteenth century,

Defoe suggested that 'in no City in the World so many People live in so little Room as at Edinburgh'; though he conceded that 'The main street . . . is the most spacious, the longest, and best inhabited street in Europe . . . the buildings are surprising both for Strength, for Beauty and for Height' (Defoe 1927: Eleventh letter).

Within 75 years, Edinburgh had gone on to become one of the great European centres of the Enlightenment – the city of David Hume, William Robertson and Robert Adam. Spurred by a desire to compensate for the city's loss of status as a national capital as a result of Scotland's 1707 Union with England and Wales, Edinburgh's large and well-educated middle-class community vigorously pursued the ideals of 'progress, prosperity, order and elegance'. One of the earliest outcomes in terms of the built environment was a high-status extension of the city towards the south-east, around George Square. This could fairly be described as a prototype new town. However, it was a later development, which became known as Edinburgh's New Town, that represents the most distinctive imprint of the neoclassical phase of the Renaissance on the city (Figure 3.4).

Following the draining of the marshy North Loch area in 1759 and the construction of a bridge (the North Bridge) in 1772, the city organized a competition in 1776 to select a plan for the New Town, won by James Craig. The New Town did not emerge precisely in accordance with Craig's plan, but it did keep to his general lines, and the whole development was enhanced by the quality of the architecture, particularly in Charlotte Square (built between 1791 and 1820), which Robert Adam had been commissioned to design. Most of the development was for wealthy families, but the New Town did incorporate a carefully planned component of residences for servants and tradesmen. Contained and integrated with the townscape are gardens, designed to take full advantage of the topography, while forming an extensive system of private and public open spaces (Edwards and Jenkins 2005).

By the beginning of the nineteenth century, extensions to the New Town had been built to the north, east and west, with crescents replacing terraces and squares as the dominant layout. Many of the later extensions, however, lacked the Georgian élan, and elegant style became a matter of unimaginative convention. Meanwhile, around 1800, there began an ambitious phase of civic development, with public buildings, monuments and statuary adding to the Georgian domestic architecture and earning Edinburgh the title the 'Athens of the North', as one of the major centres of the international Greek Revival.

Conservation has for a long time been the most distinctive aspect of urban policy in Edinburgh. The city's architectural heritage is not only an important cultural

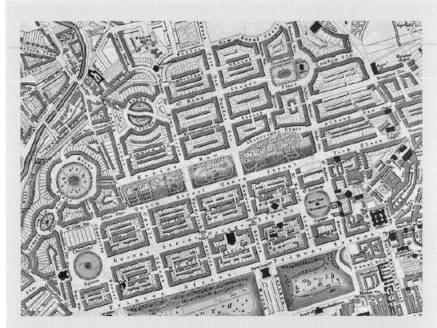

Figure 3.4 Edinburgh New Town in 1834. James Craig's design was based on a simple axial grid, with a principal thoroughfare linking two garden squares. Two other main roads were located parallel to the north and south, with two mews lanes providing access to the stables for the large homes. Completing the grid were three north–south streets. (Source: David Rumsey Map Collection, www.davidrumsey.com)

legacy but also a vital component of the tourist industry that, after office development, offers the best prospects for creating jobs in the city. In 1995, UNESCO added the New Town to its World Heritage list, citing its far-reaching influence on European urban planning and its concentration of planned ensembles of world-class neoclassical buildings, associated with renowned architects, including John and Robert Adam, Sir William Chambers and William Playfair.

Key reading

Brian Edwards and Paul Jenkins (eds) (2005) *Edinburgh: The Making of a Capital City.* Edinburgh: Edinburgh University Press.

Coping with industrialization: paternalism and philanthropy

With the onset of the industrial revolution, an entirely new set of issues emerged, both stimulating and challenging design. Figure 3.5 illustrates the chronology of the key movements and interventions of the nineteenth century. For producers, as noted in Chapter 1, innovative designs released the capacity of machinery and technology to make a profit. New technologies and new materials, in turn, allowed producers to develop new kinds of industrial buildings – vast warehouses and multistorey factories with 'fireproof' iron framing strong enough to withstand the bulky, heavy machinery that produced strong vibrations – and allowed engineers to design new sewage, water-supply and transportation systems (Goodman and Chant 1999; Roberts and Steadman 1999).

The question of what this modernizing world should look like was perplexing: what, for example, was the appropriate visual expression for a steam locomotive, a textile mill or a mass-produced teapot? Manufacturers had to create new products from first principles and to devise appropriate new forms for them. 'The crucial question for design in the 19th century', observes Deyan Sudjic, 'was how to find a language that was appropriate for machine-made objects, rather than using machines to copy forms that had evolved from handcraft' (Sudjic 2008: 17). For

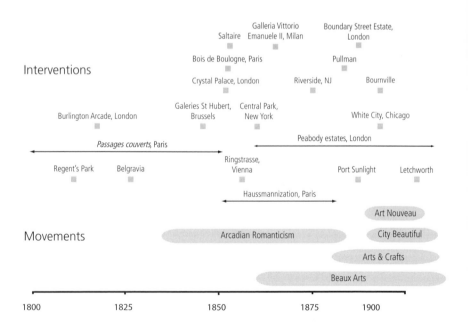

Figure 3.5 Nineteenth-century movements and interventions.

many observers, the shock of the new political economy was overwhelming, prompting them to fall back on nostalgic and idealized notions of the rural industrial past as an antidote.

Meanwhile, the productivity of factory workers, drawn in unprecedented numbers into crowded slums in rapidly industrializing cities, was threatened by poor nutrition and by endemic but life-threatening diseases such as influenza, tuberculosis, whooping-cough and scarlet fever, as well as by occasional epidemics of smallpox, cholera, typhoid and even bubonic plague. Employers and governments were constantly fearful of mob protest turning into mob rule, especially after the revolutionary events in Bohemia, the French Republic, Germany, Hungary and Switzerland in the 1840s. The respectable middle classes, meanwhile, were worried about the breakdown of moral order and the contagion of disease. In Europe, the lives of city dwellers were famously described by novelists Honoré de Balzac, Charles Dickens, Victor Hugo and Marcel Proust and catalogued by social observers Charles Booth, Friedrich Engels, Henry Mayhew and Alexandre Parent-Duchâtelet. In America, ordinary city lives were dramatized in the novels of Abraham Cahan and Stephen Crane and documented by journalist Jonathan Harrison and photographer Jacob Riis.

Little thought, however, was given to residential design or urban planning. The design antecedents of the Renaissance and Baroque eras were barely relevant to the issues faced in industrializing cities. In any case, the laissez-faire principles of the time precluded any intentional intervention in the machinations of the real estate or labour markets.

Utopianism and arcadian romanticism

It was left, therefore, for individuals and private groups to respond to the shocking new developments of the industrial revolution. Two strands of thinking emerged, often conflated with one another. One was utopian, seeking to take advantage of the productivity of the industrial revolution to create ideal physical and social environments. The other was somewhat reactionary, seeking to counter the negative consequences of rapid urbanization with arcadian settings and classical architectural motifs. The landmark utopian initiative was New Lanark, a mill village of some 2,500 people established to take advantage of the water power of the River Clyde in central Scotland. Now a UNESCO World Heritage Site and an Anchor Point of the European Route of Industrial Heritage, New Lanark was founded in 1786 by a partnership that included Robert Owen, whose progressive social philosophy made him uncomfortable with the living conditions endured by factory workers and led him to introduce factory reform and improved working practices.

Although the mills were very profitable, Owen's partners were unhappy about the cost of his welfare programmes, so he bought them out in order to be able to continue with reforms. He opened the first ever infant school in Britain in 1816 and saw the construction of not only well-designed and equipped workers' housing but also public buildings designed to improve their social and moral well-being.

Owen became an influential social reformer and his ideas on model industrial communities resonated throughout the nineteenth century. In Britain, the idea blossomed in the 1850s among certain elements of the new, self-made industrialists. The Wilson family, for example, founded their industrial village, Bromborough Pool, in Cheshire in 1853 for Price's Patent Candle Company: it was an open village of short terraces and semi-detached cottages with large gardens and considerable public open space. The Ackroyds and the Crossleys, two industrial families in Halifax, developed model housing for their workers in Ackroydon and Halifax respectively, but at a high density, with no gardens, and no concept of overall layout. The chief innovation in Ackroydon was in its financial arrangements: a scheme for workers to buy their own homes through the novel method of a building association. The most famous and most ambitious of the early model communities was Saltaire, built by Titus Salt in the Aire valley outside Bradford. Between 1853 and 1863 he built 800 houses around his new textile mill, laid out with a careful stratification of house types and sizes according to class of worker. The big innovation was his provision of institutions and amenities: almshouses, public baths, wash-houses, a sumptuous Congregationalist chapel, a Sunday School, a community institute and a park (Tarn 1973).

These were surely cases of enlightened self-interest, the increased health and contentment of their factory workers leading, it was expected, to increased productivity. Like New Lanark (where there was an Institute for the Formation of Character) they were also paternalistic, the factory owners imposing their values on their respective communities in ways that were not always appreciated by their tenants. Later in the nineteenth century, the larger and more successful examples of model industrial communities in Britain were rather less paternalistic, though they retained a strong utopian flavour. William Lever, who built Port Sunlight (Figure 3.6) on the Wirral peninsula across from Liverpool in the 1890s, regarded it as a social duty to provide his workpeople with satisfactory homes regardless of the cost, the rents paying only for local taxes plus repair and maintenance. Port Sunlight consisted of small groups of picturesque cottages laid out at a low density, giving an arcadian feel. Architecturally, it was a great success, but its strong philanthropic basis meant that it was unlikely to be widely emulated.

In contrast, George Cadbury, having moved his cocoa factory from Birmingham to a greenfield site at nearby Bournville, wanted to build a model community that

Figure 3.6 Port Sunlight, Cheshire, England. Built by William Lever to house employees in his soap factory, Port Sunlight had its first occupants by 1889 and by 1907 the planned community had 278 dwellings in a mixture of neo-traditional styles that included Old English, Flemish, and Dutch. (Photo: Time Life Pictures/Getty Images)

was financially sound to the point where, after repairs and maintenance costs had been deducted, there should still be a return of 4 per cent. In so doing, Cadbury hoped that he would be able to set local authorities an example, and encourage them to build for working classes generally. For this reason, Bournville itself was never intended primarily to house Cadbury's own workers, one of the principal intentions being to encourage a broadly mixed community.

In the United States, George Pullman, inventor of the railway sleeping car, established a company town in 1880 just south of Chicago with its own housing, shopping areas, churches, theatres, parks, hotel and library for his employees. Pullman, however, was overbearingly paternalistic: he would not allow independent newspapers, public speeches, town meetings or open discussion. His inspectors regularly entered workers' homes to inspect for cleanliness and could terminate leases at ten days' notice. When the railway business fell off in 1894, Pullman cut jobs, wages and working hours, but not rents or prices in his town. The workers rebelled, their violent strike prompting a national commission, which concluded that Pullman's paternalism was partly to blame and that the company town of Pullman was 'unAmerican'.

Intellectuals' responses to the shocking outcomes of nineteenth-century urbanization, meanwhile, were dominated by reactionary impulses. There was widespread
disdain for the values and ethics of the commerce and industry that were
transforming urban society. Industrial economic development was associated with
corruption, exploitation and moral degeneration. At the same time, there was an
abiding fear of the social and physical consequences of this exploitation and
degeneration: the alienated 'mob' and its squalid and unhealthy neighbourhoods
that were seen as a threat to physical well-being and social order. The proper
response for art and architecture, it was argued, was a celebration of Nature and
its therapeutic and uplifting properties. One especially influential English author
was John Ruskin, whose *Seven Lamps of Architecture* (1849) and *The Stones of
Venice* (1851) strongly rejected both classical architecture and the mechanization
and standardization of the industrial era. Instead, Ruskin emphasized the importance of medieval Gothic style for what he saw as its reverence for nature and
natural forms. Ruskin also campaigned for the conservation of ancient buildings,
while his advocacy of natural, organic form and his praise for the work of medieval
artisans influenced the emergence, later in the nineteenth century, of the Arts and
Crafts movement.

In the United States, and contemporary with Ruskin, design professionals were
especially influenced by the literary culture of the so-called American Renaissance.
This was rooted in the works of Ralph Waldo Emerson and Henry David Thoreau,
whose objective was to confront the future rather than turn away from it: to define
ideals appropriate to the time. Emerson had drawn, in the 1830s, on the sensibility
of European Romantics like artist John Constable. Thoreau, a disciple of
Emerson's, popularized the idea of Nature as a spiritual wellspring for city dwellers
in his book *Walden* (1854). Americans soon came to think of their relationship
with Nature and the 'great outdoors' as something distinctively 'American'. The
historian Frederick Jackson Turner (1920) advanced the influential idea that the
American 'frontier experience' was the single most significant factor in deter-
mining the American character, and it became broadly understood that 'access to
undefiled, bountiful and sublime Nature is what accounts for the virtue and special
good fortune of Americans' (Marx 1964: 12). Against this backdrop, the American
Renaissance took Nature as a fundamental spiritual wellspring, defining their
ideal as a setting in which man and nature had achieved a state of balance – what
landscape architect Leo Marx (1964) has described as a 'middle landscape' of
pastoral and picturesque settings. At the same time, the intellectuals of the period
emphasized the moral superiority of domesticity and the virtues of republicanism
and sanitary reform. This attitude led to a vision of ideal urban landscapes that
combined the morality attributed to Nature with the enriching and refining
influences of cultural, political and social institutions. Progressive intellectuals

like Andrew Jackson Downing advocated a programme of 'popular refinement' involving the creation of a whole series of institutions and settings such as public libraries, galleries, museums and parks, in order to cultivate feelings of honesty, beauty, wholesomeness, cleanliness and natural order, and bring out the best in 'ordinary' people.

Parks, in particular, became an important aspect of urban design. By bringing the working classes into contact not only with the spiritual energy of Nature but also with the enlightened manners and comportment of other classes, parks could become a kind of universal moral force, a source of democratic and fraternal feelings: 'It was believed that parks would breed a desire for beauty and order, spreading a benign, tranquilizing influence over their surroundings' (Boyer 1983: 39). The park movement was significant for several reasons. Its success consolidated the ethos of paternalism in urban policy-making and planning; it established environmental determinism as a strong element in city planning; and it brought about real change in the shape and pattern of daily life.

In London, royal deer parks like Hyde Park and the adjacent Kensington Park Gardens had been landscaped and made freely accessible to the public in the eighteenth century. In Paris, the Bois de Boulogne, a remnant of an ancient oak forest, was made into a park by Napoleon III in 1852, financed by selling building lots along the north end of the park, in Neuilly. Landscaping the park involved the creation of two lakes, connected by a waterfall, together with 35 kilometres of footpaths and 29 kilometres of riding tracks. In other cities, parks were landscaped from remnant enclaves of farmland.

In the United States the park movement was endorsed by leading intellectuals of the American Renaissance and rapidly gained momentum during the second half of the nineteenth century. The most influential figure was Frederick Law Olmsted. In 1858 he and Calvert Vaux won the competition for the design of Central Park in New York City. Their design, which was completed in 1862, included a succession of specific areas for sport, recreation and culture, all embedded in a picturesque landscape. The park was integrated with the city by means of four avenues laid out with an elaborate system of independent traffic lanes, bridges and underpasses that were designed not to interrupt the continuity of the landscape. The result was widely acclaimed, and by the turn of the century, large parks had been established in Baltimore, Boston, Chicago, Cleveland, New York, Philadelphia, St Louis, San Francisco and Washington, DC. New York had eleven parks of 100 acres or more. Boston had initiated the first planned metropolitan system of parks: an 'emerald necklace' of parks and parkway links. Olmsted himself went on to design park projects in other cities (including Boston, Brooklyn, Buffalo, Chicago, Detroit, Milwaukee, Newark, Philadelphia and San Francisco) and campuses for the University of California at Berkeley and Columbia University in New York. Smaller parks proliferated in cities everywhere: ornamental parks, zoological parks, parks for strolling, boating, lunching, skating and team sports, waterfront parks, downtown mini-parks and neighbourhood parks.

Reform and philanthropy

While the interventions of paternalistic philanthropists and the growth of the park movement were significant antecedents, in retrospect, of later developments in urban design and planning, they made relatively little impact on the economic, social and environmental problems of cities that were growing at unprecedented rates and without any significant regulation. The first real focus of concern had come from the efforts of one individual in Britain: Edwin Chadwick, First Secretary of the Poor Law Board. His campaigning led to the Royal Commission on the Health of Towns, whose shocking report (in 1842) encouraged progressive-minded liberals to form themselves into voluntary associations such as the Health of Towns Association, the Association for Promoting Cleanliness Amongst the Poor, the Society for Improving the Condition of the Labouring Classes, and the Metropolitan Association for Improving the Dwellings of the Industrious Classes. At first these associations limited themselves to discussions and the passing of resolutions, but soon they began to lead by example in constructing demonstration projects that sought to show how decent housing could be built for working-class households with affordable rents that still yielded a (modest) profit.

The problem that was that the profit margins on such projects were far too modest to attract much serious attention from investors or developers, and meanwhile the urban crisis intensified. By the 1860s, the liberal response in Britain was dominated by wealthy philanthropists such as the Guinness family and American banker and diplomat George Peabody, who were willing and able to build worker housing that yielded profits that were, for the period, strikingly low: typically around 5 per cent (Tarn 1973). On this basis, the Peabody Trust alone built more than 20,000 dwellings in London between 1862 and 1890, some of which are still in operation (Figure 3.7). Peabody was part of a circle of reformers that included Lord Shaftesbury, William Cobbett and Charles Dickens. He pioneered social housing with such unheard-of amenities as separate laundry rooms and space for children to play. The first Peabody estate, in Spitalfields, was opened in 1864. Since then, the Peabody Trust has become a modern housing association, with new housing (Figure 3.8) as well as renovated nineteenth-century properties, altogether housing almost 50,000 people. Such philanthropy did not affect the poorest of the poor (who could not afford the rents), nor did it make significant inroads to the ever-worsening conditions of industrial cities. However, it did help to gain some

Figure 3.7 Peabody Estate, Wild Street, Westminster, London. The Peabody Trust was founded in 1862 by an American merchant banker, George Peabody, to provide housing for people in need in London. Built in 1882, it was modernized in the 1960s. (Photo: author)

Figure 3.8 Peabody Estate, Nile Street, Hackney, London. Built in 2005, this development provides a total of 175 homes, 128 of them affordable and designed for key workers and those on low or intermediate incomes. The sale of the private flats has cross-subsidized the affordable properties of the development. There are also communal and roof gardens as well as a youth centre within the development. (Photo: author)

attention within polite society on the issue of how to address the conundrum of industrial cities: sustaining environments in which investors could make profit in both the labour market and the housing market without leaving large numbers of households so poor that their slum neighbourhoods threatened the health and security of everyone else.

It took a steady succession of epidemics and public health crises to bring the issues to the top of the public (middle-class) agenda. In Britain, the 'problem of towns' was debated by learned societies and parliament, prompting a Royal Commission that in turn resulted in the landmark Public Health Act 1875. This had two major consequences: provision for local authorities to enact bye-laws about housing design and construction, and to develop policies directed toward compulsory purchase and slum clearance. In the 1880s the continuing efforts of liberal reformers to expose the plight of the urban poor brought the issue to the popular press, and with that there began a more widespread acceptance for public intervention in urban development. Particularly important were the Reverend Andrew Mearns, whose graphic descriptions of slum neighbourhoods were published in *The Bitter Cry of Outcast London: An Inquiry into the Condition of the Abject Poor* in 1883, and Charles Booth, whose careful statistical compilations and ethnographic studies in London were published in *The Life and Labour of the People* in 1889. Reflecting popular acknowledgement of the need for public intervention, and recognizing that neither private developers nor philanthropists could derive sufficient profit from redeveloping the worst slums, the central government granted permission for London County Council to build public housing estates in slum clearance areas. The world's first public housing project, the Boundary Street Estate, was built between 1880 and 1890 on the old Friars Mount slum in Bethnal Green, in the East End of the city.

Design for consumption

For the affluent middle classes, the industrial revolution brought new levels of material consumption, and new settings in which to purchase and consume them. Those settings, in turn, led to new forms of sociability that created a new fashion- and design-conscious atmosphere in the downtown districts of larger cities. Developers took advantage of new materials like plate glass and new cast iron and steel construction techniques in their new railway stations, hotels, cafés, restaurants and department stores. Such places embodied modern progress, the stores brimming with wave after wave of new commodities from the flood of industrial mass production and colonial exploitation, the stations, hotels and cafés abuzz with a new feeling of dynamism.

It was in Paris, the capital of modernity in the nineteenth century, that the precursors of this change took form (see Case Study 3.2). Elsewhere, the innovation of arcades was copied in a variety of different formats. The Burlington Arcade, off Picadilly, in London, for example, was built in 1819 as an exclusive, high-end retail setting.

Case Study 3.2 **The *passages* of Paris**

The *passages* of Paris offered unheard-of amenities to the emerging class of bourgeois consumers in the first half of the nineteenth century. With specialized shops offering a broad range of goods and services in a contained space, gas lighting, warmth and shelter from rain and mud, and cafés and restaurants where shoppers could rest and observe fellow lingerers, they were magnetic in their attraction. *The Illustrated Guide to Paris* from 1852 neatly summarized the appeal:

> In speaking of the inner boulevards, we have made mention again and again of the arcades which open onto them. These arcades, a recent invention of industrial luxury, are glass-roofed, marble-panelled corridors extending through whole blocks of buildings, whose owners have joined together for such enterprises. Lining both sides of these corridors, which get their light from above, are the most elegant shops, so that the arcade is a city, a world in miniature, in which customers will find everything they need.

The Passage des Panoramas, off the boulevard Montmartre, is one of the oldest (built in 1823) and one of the first public places in Paris to have gas lighting. It got its name from the painted panoramas that were once projected on twin rotundas inside the passage. The Galerie Véro-Dodat (Figure 3.9), opened by two pork butchers in 1826, boasted a black-and-white geometric marble floor, ball lamps and painted Corinthian columns. The Galerie Vivienne, between rue Vivienne and rue des Petits Champs, opened to the public in 1826 with a range of expensive stores and fashionable cafés. The Passage Jouffroy (1847) was the first heated arcade and the first built entirely of iron and glass. Altogether, some 150 *passages* were built in Paris during the first half of the nineteenth century. A few of them specialized in a single commodity – such as fish in the Passage du Saumon, and lithographs in Passage du Caire, but most housed a mix of milliners, hosiers, haberdashers, tailors, bookshops, bootmakers, wine merchants, caricaturists, print shops and cafés. Fewer than 20 of them remain today, but most of these have been restored and revitalized, like the Galerie Vivienne (Figure 3.10).

Walter Benjamin, writing about the *passages*, saw their impact as the moment of transformation from a culture of production to one of consumption. Capitalism, he

Figure 3.9 Galerie Véro-Dodat, Paris. Restored in the 1980s after a century
of neglect. (Photo: author)

Figure 3.10 Galerie Vivienne, Paris. Opened in 1826 as a luxury *passage*, it was immediately popular. (Photo: author)

argued, endowed objects with the means to express collective dreams. This drew him to particular urban architectural forms, 'dream houses of the collective' such as the *passages*, railway stations, department stores and wax museums. As historian Vanessa Schwartz notes:

> If *The Arcades Project* suggests anything, it is that modernity cannot be conceived outside the context of the city, which provided an arena for the circulation of bodies and goods, the exchange of glances, and the exercise of consumerism.
>
> (Schwartz 2001: 1733)

By mid-century, ambitious developers had expanded the scale of the idea to the galleria, a spacious indoor court rather than a narrow passageway. The first example was the Galeries Royales Saint-Hubert in Brussels (1846); other examples are the Galleria Vittorio Emanuele II in Milan (1865–1877, Figure 3.11), the Barton Arcade, off St Ann's Square and Deansgate, Manchester (1874) and the Galleria Umberto I in Naples (1887–1891): precursors of the twentieth-century shopping mall.

Figure 3.11 Galleria Vittorio Emanuele II, Milan. Built by Giuseppe Mengoni between 1865 and 1877, the Galleria connects two of Milan's most famous landmarks: the Duomo and the Teatro Alla Scala. (Photo: author)

The successors to the *passages* and gallerias were department stores: larger, more spectacular and more respectable, with no dubious nightlife to contend with. Mark Jayne (2006) writes that department stores

> provided the material means for the middle class in particular to stake out their cultural identity. Of importance was the respectable presentation of service staff and surroundings, which included an abundance of marble, carpets and ornaments – the refined sophistication of galleries with a luxurious, almost aristocratic, ambience. Department stores sought to replicate the design and décor of theatres and art galleries with bright lighting, and grandly designed halls. It is through their construction in terms of their relative safety and organization that department stores became an important female public space, a realm where commodities and gender became intertwined.
>
> (Jayne 2006: 43)

The introduction of electric lighting toward the end of the nineteenth century allowed for even more dazzling displays, while the introduction of electric street lighting made entire shopping districts much safer and more attractive.

The impact of these settings on people's comportment and attitudes was noted in the 1860s by the French poet and literary critic Charles Baudelaire (1965). Somewhat the dandy himself, Baudelaire wrote about the decadence of city life, describing the city as an 'urban phantasmagoria', the locus of modernity with its fleeting, ephemeral and contingent experiences. The lives on display in the *passages* and gallerias also fascinated writers like Honoré de Balzac (1837), who wrote about the booksellers and milliners who hawked their wares in the Galerie de Bois, and Emile Zola (1867), who used a grimy, sinister passage as the claustrophobic setting for his macabre *Thérèse Raquin*. A new social type, the *flâneur*, was also identified with the new commercial settings of the modern city. Famously documented by Walter Benjamin in *The Arcades Project* (1999), the *flâneur* was a bourgeois male observer of the patterns and rhythms of city life. The *flâneur* sought out the sights of the city and its tumultuous crowd, on the lookout for the new, exciting and unfamiliar; window-shopping, wanting to see and be seen, while seemingly remaining aloof and detached. The *flâneur* is an interesting social type because it underscores the centrality of movement and variety in modern urban life: the stroller is constantly exposed to new streams of experience and develops new perceptions as he moves through the crowds and urban landscape.

The city of good intentions: imposing order, safety and efficiency

As cities grew larger and more complex, national rulers and city leaders looked to urban design to impose order, safety, and efficiency, as well as to symbolize the new seats of power and authority. One of the most important early precedents was set in Paris by Napoleon III, who presided over a comprehensive programme of urban redevelopment and monumental urban design. The work was carried out by Baron Georges Haussmann between 1853 and 1870 (see Case Study 1.1). Haussmann demolished large sections of old Paris to make way for broad, tree-lined avenues, with numerous public open spaces and monuments (Figure 3.12). In doing so, he made the city not only more efficient (wide boulevards meant better flows of traffic) and a better place to live (parks and gardens allowed more fresh air and sunlight in a crowded city and were held to be a 'civilizing' influence) but also safer from revolutionary politics (wide boulevards were hard to barricade; monuments and statues helped to instill a sense of pride and identity). The result was a powerful visual sense of modernity. As Guy Julier observes (2000), the rebuilding of Paris

> provided sweeping vistas of the cityscape so that each walk included dramatic visual effects. The boulevards became stage sets for an audience of the street-side cafés. The street provided a scenery for the middle-class stroller to see and be seen – the deprived and the destitute having been removed from the action altogether. Here, the anonymity of the crowd gave both the security of private life and the stimulation of public action.

> (Julier 2000: 120)

Figure 3.12 The Boulevard des Italiens, Paris, c. 1890. One of the four 'grands boulevards' of central Paris and part of Haussmann's modernization of the city's infrastructure. The new street quickly became very fashionable, with several famous cafés: Café de Paris, Café Tortoni, Café Frascati, Café Français and Maison Dorée. (Credit: Hulton Archive/Getty Images)

It was a city designed for looking, and for being looked at. The dominant architectural style for new buildings was the Beaux Arts style, which took its name from L'Ecole des Beaux Arts in Paris. In this school, architects were trained to draw on examples from Imperial Roman architecture, the Italian Renaissance and French and Italian Baroque styles, synthesizing them in designs for new buildings for the industrial age. The idea was that the new buildings would blend artfully with older palaces, cathedrals and civic buildings.

In Vienna, Emperor Franz Joseph I insisted on the modernization of the city in a way that was reminiscent of the Grand Designs of the Renaissance. The central element was the Ringstrasse, a broad, curvilinear boulevard commissioned in 1857 to replace the thirteenth-century city ramparts. The Ringstrasse was intended to be a showcase for imperial Habsburg grandeur, bordered with impressive public buildings (including the Parliament building, the town hall, Vienna State Opera, the Burgtheater and the Stock Exchange) that stressed secular culture and the new constitutional government; and laid out with a generous amount of parks, gardens, monuments and statues. The first public buildings were designed in neoclassical styles, consonant with Franz Joseph's idea of imperial grandeur. By the time the Ringstrasse was completely built, on the eve of the First World War, it was lined with public buildings and upscale residences in a variety of styles, including the fashionable and forward-looking Jugendstil (see p. 95).

The City Beautiful

In the United States, architects and developers seized on the Beaux Arts style as a way of resolving confusion and uncertainty about how to build in ways appropriate to urbanization in cities that had neither precedent nor legacy. The idiom was showcased by Daniel Burnham's neoclassical architecture for the World's Columbian Exposition in Chicago in 1893. The temporary structures of the Exposition showed what might be done, and before long Beaux Arts buildings were popping up across the United States, in suburbs and small towns as well as in city centres.

Almost at once the Beaux Arts ideal fused with the reformist ideology of the Progressive Era, giving rise to the City Beautiful movement (Wilson 1989; Chambers 1992). The thrust of the movement was decisively toward the role of the built environment as an uplifting and civilizing influence, with Beaux Arts neoclassical buildings and matching statuary, monuments and triumphal arches – all, if possible, laid out like Burnham's White City at the World's Columbian Exposition (Figure 1.3), with uniform building heights and imposing avenues with dramatic perspectives.

This neoclassical vocabulary brought a deliberately strong element of conservatism to the whole enterprise. The symbols and motifs of Beaux Arts architecture and monuments not only suggested a link with the great European cities of the past but also helped to legitimize the United States' Anglo-Saxon ruling classes and institutions at a time of massive immigration and profound socio-economic change. At the same time, the broad boulevards, malls and radiating road networks that framed City Beautiful projects were welcomed by civic boosters as providing an orderly physical framework for economic development and by landowners whose property values escalated in anticipation of the implied redevelopment of large tracts of central city land. The City Beautiful movement therefore flourished – albeit briefly – because it allowed private enterprise to function more efficiently while symbolizing a noble idealism that was endorsed by the pedigree of Beaux

Arts neoclassicism. Over and beyond the good intentions of its leading figures and technicians, 'the City Beautiful Movement perfectly fulfilled its true function of matching maximum planning with maximum speculation' (Tafuri and Co 1986: 40).

In 1901, Burnham collaborated with several others (including Frederick Law Olmsted, Jr) on the McMillan Plan for Washington, DC (named after Senator James McMillan, Chairman of the Senate Committee on the District of Columbia). The purpose of the McMillan Plan was to rescue the Mall area from the neglected and unfinished framework derived from Pierre Charles L'Enfant's original plan of 1791. The centrepiece of the new plan was the redeveloped Mall and Federal Triangle, with neoclassical buildings along the Mall, a terminal memorial (the Lincoln Memorial), a pantheon (the Jefferson Monument), the Memorial Bridge and a water basin. Although the scheme was not completed until 1922, the plans and sketches provided enough publicity to ensure the immediate future of the City Beautiful movement. Burnham went on to draw up plans for Cleveland (in 1902), San Francisco (1905) and Chicago (1909) before his death in 1912. These plans were also very influential, though little of the San Francisco plan was actually realized.

The success of the movement helped to foster several key developments in architecture and urban design. The first was the development of city planning commissions and the resultant awareness of a need for technical experts to prepare plans. Burnham himself famously advised cities to 'Make no little plans'. His own plan for Chicago, published in 1909, was a magnificently illustrated volume that was truly metropolitan in scope, with proposals for a system of regional ring roads connected to downtown Chicago by a series of radial highways and parkways, for over 60,000 acres of monumental parkland, and for a series of parks, marinas and developments along the lakeshore. It was a scheme for expansion, a framework for speculation, a design to embellish the city, to flatter its leaders and to impress its inhabitants. It did nothing, however, to address the fundamental problems of slum housing and social malaise.

The second was professionalization. In 1897, for example, the American Park and Outdoor Art Association was founded, followed by the American Society of Landscape Architects (1899) and the National Playground Association in 1906; courses in landscape architecture were introduced at Harvard University in 1900. The third development was an intensification of discourse and critical analysis on the aesthetics of the built environment. The City Beautiful movement had brought architecture and urban design to the forefront of public debate, provoking an intense critique of bourgeois art, architecture and urban design that, in turn, helped energize the emerging design professions.

It is clear, looking back, that the City Beautiful movement was an explicit and rather authoritarian attempt to create moral and social order in the face of urbanization processes that seemed to threaten disorder and instability. Its success did not last long, however, because, as in Europe, the dynamics of urbanization were changing. Trams and suburban railways had begun to turn cities inside out and, by the time of Burnham's death in 1912, motor cars were beginning to make their mark. In this new context, monumentality was seen as impractical, while the movement's total lack of concern for housing was seen as a major shortcoming.

Restorative utopias: garden cities and new towns

The question of designing better residential settings prompted a resurgence of idealism, arcadianism and utopianism. In the tradition of the American Renaissance, Olmsted and Vaux had already seen planned suburban development as a chance to provide the benefits of city amenities without the congestion, tumult, noise, crime and vice together with the benefits of country life without the inconvenience, isolation and lack of amenities. Their first project, in 1869, was Riverside, a Romantic-styled railway suburb 9 miles from downtown Chicago. Lushly landscaped in a park-like setting, Riverside's homes were generously proportioned, securing privacy and status for their affluent residents. Its communal open spaces were expected to be arenas for the emergence of a progressive, reformed community life. The enchantment of Riverside was underpinned by the existence of a ready-made commuter system – the railway – and soon other garden suburbs, including Highland Park, Lake Forest and Winnetka, sprang up around Chicago's local railway network. Over the course of the late nineteenth and early twentieth centuries, similar 'picturesque enclaves' emerged around other large cities, catering to upper-middle-class railway commuters (Hayden 2003).

Meanwhile, the broader reform movements of the late nineteenth century prompted others to explore a more ambitious idea altogether: the concept of garden cities that would cater not only to the upper-middle classes but also to the full spectrum of society, with jobs and civic amenities as well as homes: 'restorative utopias' (Pinder 2005). In England, Ebenezer Howard distilled and codified the garden city idea in his famous book *To-morrow: A Peaceful Path to Real Reform*, published in 1898 and reissued in 1902 as *Garden Cities of To-morrow*.

Howard's rationale and plans drew heavily on philanthropic ideas of the time, together with the aesthetic principles of William Ruskin, the communitarianism of Charles Fourier, the socialist anarchism of Peter Kropotkin and the socialist ideals of William Morris (Fishman 1987). Echoing Olmsted and Vaux's rationale for garden suburbs, he portrayed the garden city ideal as providing the possibility of combining the best of city life (jobs, higher wages, civic amenities, social inter-action, etc.) with the best of life in the countryside (clean air, natural beauty, open space) – while avoiding the downside of both (the congestion and pollution of cities; the limited employment opportunities and poor infrastructure of rural areas).

Howard's ideal plan limited each garden city to 30,000 people on 6,000 acres (about 9 square miles). The built-up area was to be about 1,000 acres (1½ miles in diameter), at the centre of which were public gardens surrounded by civic buildings – the town hall, courthouse, library, museums and hospital – easily accessible by radial boulevards. This walking city would contain a concentric zone of housing and also a separate but accessible zone of factories and warehouses connected by a circumferential rail line. The built-up area was to be surrounded by a permanent green belt that restricted urban sprawl and offered recreational opportunities while at the same time protecting agriculture.

Howard invested his own money in the concept, co-founding the Garden City Pioneer Co. Ltd, which developed Letchworth (north of London), the first full garden city (Figure 3.13). Using Howard's schematic plans, the company laid out

Figure 3.13 Letchworth, England. Founded in 1903 by Ebenezer Howard, with Barry Parker and Raymond Unwin as architects, Letchworth was the first garden city. In keeping with Garden City ideals, only one tree was felled during the entire initial construction phase of the town. (Photo: Topical Press Agency/Getty Images)

roads, parks and factory sites and invited private developers to build housing (within carefully regulated standards) on prepared sites. The scheme was supported by liberal reformers because it involved utopian ideals that invoked pastoralism and social order, by practical reformers because it involved land use control and centralized direction, and by conservatives because it gave private business more scope to develop real estate. Other garden cities were built, including Hellerau near Dresden, Germany, Floreal near Brussels, Belgium, and Tiepolo, in Italy. Although Howard's logic was based on stand-alone settlements, the appeal of his urban design principles also gave rise to garden suburbs. Examples include Wythenshawe, a suburb of Manchester, England, Hampstead Garden Suburb in northwest London, Römerstadt near Frankfurt, Germany, Hirzbrunnen in Basel, Switzerland, and Milanino outside Milan, Italy.

The preferred aesthetic for many of these projects drew heavily on the Arts and Crafts movement, which flourished between 1880 and 1910. Prompted by a reaction to the dehumanizing effects of industrialization and inspired by a romantic idealization of the craftsman taking pride in his personal handiwork, the Arts and Crafts movement was led by William Morris, an artist, writer, architect, and furniture and textile designer. Morris made design a kind of moral crusade, and is famous for a phrase in his 1882 book, *Hopes and Fears for Art*: 'Have nothing in your home that you do not know to be useful and believe to be beautiful.'

The aesthetics of the Arts and Crafts movement were influenced by the writings of John Ruskin and Augustus Pugin and their advocacy of Gothic revival styles. Natural materials were preferred, and the dominant design motifs were stylized flowers, allegories from the Bible and literature, upside-down hearts, Celtic patterns and the minimalism of Japanese art. Members of the movement sought to copy the medieval system of trades and guilds, setting up their own companies to sell their goods.

Unfortunately, as with the garden cities and garden suburbs themselves, their products were not priced within reach of the great mass of the working classes. Staying true to craft techniques meant that the products of Arts and Crafts designers were well beyond the reach of the working classes whose interests and well-being they championed. With few exceptions (such as Christopher Dresser, who was ready to work with industry, new technology and new materials, and is regarded by some as the first industrial designer), Arts and Crafts designers rejected the mass production methods that could produce cheap, affordable products because they did not want to turn independent craftsmen into 'wage slaves'. This dilemma was not lost on William Morris, who asked 'who will free me from pandering to the luxury of the swinish rich' (Sudjic 2008: 25). In the longer term, the principal impact of the Arts and Crafts movement was on the intellectual and artistic community itself, the emphasis on craft and nature carrying over to the Art Nouveau movement and to the early pioneers of German Modernism.

The practical successes of the European garden cities resonated resoundingly with intellectuals and developers in the United States. The Russell Sage Foundation developed the first garden city in the United States, Forest Hills Gardens, in Queens, 9 miles from Manhattan, in 1911. Designed by Grosvenor Atterbury and Olmsted's son, Frederick Law Jr, Forest Hills Gardens contained a kitsch-like mix of housing laid out with a distinctive neighbourhood structure that convinced one of its residents, Clarence Perry, that the layout of a project could, if handled correctly, foster 'neighbourhood spirit'. Perry developed the concept of the neighbourhood unit, defined by the catchment area of an elementary school, focused on the school itself, local stores and a central community space, and bounded by arterial streets wide enough to handle through traffic (Biddulph 2000). As Peter Hall (2002) points out, Perry saw the neighbourhood unit as an opportunity for social engineering that would assist in nation-building and the assimilation of immigrants, but the idea of neighbourhood spirit also went down well with communitarians, and the idea of handling traffic went down well with planners who were beginning to grapple with the implications of the spread of car ownership (Hall 2002).

During the 1920s and 1930s Perry's neighbourhood unit concept was adopted and developed by others – notably Clarence Stein and Henry Wright, in landmark developments that included Sunnyside Gardens (New York), Radburn (New Jersey), Chatham Village (Pittsburgh) and Baldwin Hills Village (Los Angeles). John Nolen (1927), drawing on the ideals of the American Renaissance, sought to revive the association of physical design with the civic ideals in his book *New Towns for Old*. Nolen advocated the pursuit of what he called the 'Greek Ideal' – rationality, the study of nature, and celebration of life – through urban design. Meanwhile, garden city ideals were being adopted and sponsored by the Regional Planning Association of America, whose evangelical spokesperson-in-chief was historian and critic Lewis Mumford. These ideas were to exercise enormous influence on American planning practice and planning ideology from the 1930s onwards, not least through the legacy of Franklin Delano Roosevelt's New Deal programmes and, more recently, the New Urbanism movement (see Chapter 5).

Toward modern design

The period between 1880 and 1930 was pivotal for cities and design. A dazzling succession of innovations and new technologies – street and interior lighting, tram systems, internal combustion engines, pneumatic tyres, aircraft, steam-turbine engines for oceanic shipping, telegraph, radio, the magnetic recording of sound, movies, synthetic fibres, X-rays, iron-case skyscraper frameworks – shook people's sensibilities, turned cities inside out and upside down, and provoked the beginnings of modern design movements. As Robert Hughes (1980) observed in his masterful book on modern art, *The Shock of the New*:

> In 1913 the French writer Charles Peguy remarked that 'the world has changed less since the time of Jesus Christ than it has in the last thirty years'. He was speaking of all the conditions of Western capitalist society: its idea of itself, its sense of history, its beliefs, pieties, and modes of production – and its art. . . . Between 1880 and 1930, one of the supreme cultural experiments in the history of the world was enacted in Europe and America.
>
> (Hughes 1980: 1)

Two briefly flourishing movements, in particular, were key to the transition from the nineteenth to the twentieth century in art, thought, and society, the precursors of the Modernism that was to dominate design from the 1930s onwards: Art Nouveau, or Jugendstil (1890–1905) and Art Deco (1925–1939).

Art Nouveau is associated with the architecture, furniture and jewellery of Charles Rennie Mackintosh, the posters of Alphonse Mucha, the paintings of Gustav Klimt, the illustrations of Aubrey Beardsley, the glass and jewellery of René Lalique, the lighting of Louis Comfort Tiffany, the architecture of Antonio Gaudí, Victor Horta and Otto Wagner, and the Paris Métro entrances of Hector Guimard (Figure 3.14).

Figure 3.14 Paris Métro entrance. Hector Guimard's Art Nouveau design for the new metro system, installed between 1899 and 1905, was in deliberate opposition to the ruling taste of French classical culture. (Photo: Stefano Bianchetti/Corbis)

Art Nouveau sought to be progressive, and it developed at a time when people had become increasingly aware of the beneficent impacts of the machine age, even beginning to be excited by it, rather than repelled by the steam, smoke, and squalor of mid-century industry. As a result, Art Nouveau was oriented toward mass markets and, unlike the Arts and Crafts movement, comfortable with mass production. Nevertheless, Art Nouveau shared the same belief in quality goods and fine craftsmanship as the Arts and Crafts movement. It was characterized by flowing curvilinear forms, by organic shapes, (especially floral and other plant-inspired motifs – stylized flowers, leaves, roots, buds and seedpods, for example), by the female form in pre-Raphaelite poses with long, flowing hair, and by the use of exotic woods, marquetry, iridescent glass, silver and semi-precious stones.

Some of the motifs drawn from nature carried over from Art Nouveau to Art Deco but the flowing organic shapes were replaced by geometric shapes and streamlined styling. Art Deco had no strong philosophical or political roots or intentions: it was a purely decorative, hedonistic movement that was directly influenced by the events and technologies of the time. Aviation, electric lighting, the radio, ocean liners and skyscrapers inspired streamlined styling and generated a spin-off genre, Streamline Moderne, or simply Streamline. Cubism and experimental art inspired dramatic geometric forms, chevron patterns, and zigzag and sunburst designs. Early Hollywood inspired the use of luxurious fabrics, specialized lighting and mirrors. And the stunning archeological discovery in Egypt in 1922 of the tomb of King Tutankhamun ignited public interest in the arts of 'primitive' societies and inspired the deployment of patterns and icons taken from the Far East, ancient Greece and Rome, Africa, India, and Mayan and Aztec cultures as well as ancient Egypt.

Art Deco is associated in particular with the furniture of Eileen Gray, the jewellery of Raymond Templier, the chinaware of Clarice Cliff, the architecture of Raymond Hood (designer of the Radio City Music Hall auditorium and foyer and the RCA Building at Rockefeller Center in New York City), William Van Alen (designer of New York's Chrysler Building – Figure 3.15) and the graphic design and industrial design of Raymond Loewy (designer of streamlined locomotives, the Shell logo, the Lucy Strike cigarette pack and Sears refrigerators). More generally, Art Deco design was deployed widely during the 1920s and early 1930s in the interior design of ocean liners, cinemas and railway stations, in branding and packaging, and in domestic soft furnishings. By the late 1930s Art Deco had declined in popularity but its emphasis on functionality and its of-the-moment sensibility helped to pave the way for the Modernism of the twentieth century. Interestingly, when Modernism faltered for a while in the 1980s and 1990s, postmodern design – another hedonistic movement – drew heavily on Art Deco: as, for example, in

Figure 3.15
The Chrysler Building,
New York City.
A classic Art Deco
skyscraper designed by
architect William Van
Alen and completed in
1930. (Photo: Joseph
Sohm/Corbis)

Memphis furniture, and the architecture and product design of Michael Graves and Aldo Rossi (see Chapter 5).

Summary

The driving forces behind the key antecedents of urban design and planning have ranged from despotic powers to utopian idealists, and from democratic govern-ments to private developers. The early origins of modern Western urban planning and design can be traced to the Renaissance and Baroque periods in Europe, when artists and intellectuals dreamed of ideal cities, and rich and powerful regimes used urban design to produce extravagant symbolizations of wealth, power and destiny. But with the onset of the industrial revolution, an entirely new set of issues emerged, both stimulating and challenging design. New technologies and new materials allowed producers to develop new kinds of industrial buildings and allowed engineers to design new sewage, water-supply and transportation systems. For the affluent middle classes, the industrial revolution brought new levels of material consumption, and new settings in which to purchase and consume them: *passages*, gallerias, department stores and new, suburban residential settings. These changes, in turn, led to new forms of sociability that created a new fashion- and design-conscious atmosphere in the downtown districts of larger cities.

It was left to a few individuals and private groups to respond to the unwanted new developments of the industrial revolution. Two main strands of thinking emerged. One was utopian, seeking to take advantage of the productivity of the industrial revolution to create ideal physical and social environments. The other was reactionary, seeking to counter the negative consequences of rapid urbanization with classical architectural motifs and arcadian settings. As cities grew larger and more complex, national governments and city leaders looked to urban design to impose order, safety and efficiency, as well as to symbolize their new seats of power and authority. Haussmann's restructuring of Paris provided an important precedent and example, as did the City Beautiful movement in the United States. The broader reform movements of the late nineteenth century prompted others to explore the concept of garden cities: 'restorative utopias' that would cater not only to the upper-middle classes but also to the full spectrum of society, with jobs and civic amenities as well as homes.

Meanwhile, the new technologies and rapidly changing social context of the industrial era gave rise to new artistic and cultural responses that led to the emergence of modern design. Between 1880 and 1930 a succession of innovations and new technologies shook people's sensibilities, turned cities inside out and upside down, and provoked the beginnings of modern design movements.

Further reading

Peter Hall (2002) *Cities of Tomorrow: An Intellectual History of Urban Planning and Design in the Twentieth Century*, 3rd edn. Oxford: Blackwell. The classic history of modern urban planning. Hall provides a critical history of planning in theory and practice as well as of the social and economic problems and opportunities that gave rise to it.

David Harvey (2000) *Spaces of Hope*. Berkeley, CA: University of California Press. Utopian movements have for centuries tried to construct a just society. Harvey looks at their history to ask why they failed and what the ideas behind them might still have to offer.

Anthony Morris (1994) *History of Urban Form: Before the Industrial Revolutions*, 3rd edn. London: Longman. A detailed history of urban development up to the industrial revolution. It contrasts 'unplanned' cities that grew organically with 'planned' cities that were shaped in response to formal interventions.

John Tarn (1973) *Five Per Cent Philanthropy*. Cambridge: Cambridge University Press. An illustrated social and architectural history of urban reform movements in Britain.

4 **The city redesigned**

Modernity, efficiency and equity

This chapter traces the emergence of urban design and planning and its relationship to Modernism. It shows how urban design and planning developed in response to the need to address both the efficient functioning of cities as settings for economic development and the amelioration and mitigation of the unwanted side effects of development. Over the course of the first half of the twentieth century, professionalized planning and urban design took on more and more responsibility for ensuring the efficient management of urban land, infrastructure and services, drawing all the time on an idealistic and progressive frame of reference and the associated design ideology of Modernism. The 'golden age' of urban planning, from 1950 to the mid-1970s, was characterized by a proliferation of government programmes for housing, urban renewal, land use zoning, transportation planning, environmental quality and comprehensive planning projects. By the late 1970s, however, changing economic circumstances, combined with disillusionment about the effectiveness of these initiatives, led to widespread critiques of planned modernity and the emergence of a very different set of political and cultural sensibilities.

The idea that the unwanted side effects of urbanization processes can be mitigated or prevented by the design of buildings, streets, neighbourhoods and even entire towns and cities has a long history. It is an idea that has been given shape by remarkably few key concepts, all stemming from the rationalism of modernity (Madanipour 2007). These concepts have influenced successive phases of urban development in the cause of the economic efficiency, social stability and the enchantment of consumers that is necessary to urban order and to sustained capital accumulation and circulation. The common denominators, in the evolving canon of architecture, planning and urban studies, have been overwhelmingly prescriptive and deterministic, often involving a raw environmental determinism and the privileging of spatial form over social process. In some cases (such as the urban renewal programmes of the 1960s and 1970s), the evangelism and environmental determinism of planners and urban designers has led them to pursue 'bureaucratic offensives' with negative consequences.

Urban design professionals find themselves in a pivotal but problematic position. They are inextricably implicated in the successes and failures of urban development, yet their roles are often ambiguous and ambivalent. They cast themselves as stewards of the built environment, guardians of the public interest and champions of the aesthetics of cityscapes, and they are perceived that way by large sections of the public. But at the same time they are servants of their clients and employers. They do not legislate policies, nor do they, typically, invest capital in urban development. Yet their influence on outcomes can be critical. This Janus-faced condition of the urban design professions is founded on a deep paradox that persists in the make-up of contemporary planning and urban design. The paradox is this: although urbanization was the vehicle that industrial capitalism needed in order to marshal goods and labour efficiently, it created dangerous conditions that threatened public health and resulted in crowded settings within which the losers and the exploited might organize, consolidate and rebel. Urban design and planning, as a response to this paradox, were born as hybrid creatures, dedicated at one level to utopian ideals of humanistic and sanitary reform, but charged on another with ensuring the efficient management of urban land, infrastructure and services, as well as fostering and maintaining 'balanced' and stable communities.

As such, architecture, planning and urban design can be construed as key to the internal survival mechanisms of capitalism, channelling the energy of opposing social forces into the defence of the dominant order, helping to propagate its own goals and values as the legitimate ones (Knox and Cullen 1981). Their practitioners can be seen as Weberian 'ideal types': key actors in the development process, with distinctive values and ideals that are routinely transcribed into the built environment through their day-to-day work as well as through their more visionary prescriptions. Planning and urban design rarely determine large-scale form; rather, urban form follows market forces, and planners 'smooth' the rough spots. Many of the basic decisions about urban development are taken by others. As Jonathan Barnett (1982) points out, the form of cities

> is usually unintentional, but it is not accidental. It is the product of decisions made for single, separate purposes, whose interrelationships and side effects have not been fully considered. The design of cities has been determined by engineers, surveyors, lawyers, and investors, each making individual, rational decisions for rational reasons.
>
> (Barnett 1982: 9)

The point, though, is that planners and urban designers do not need to do anything dramatic in order to perform their function as an internal survival mechanism for capital accumulation.

It would be naive, however, to seek to trace a one-way relationship between the development of modern capitalism, the emergence of the modern planning movement, and the design and regulation of settings conducive to the profitability, stability, and social reproduction. Indeed, one would expect to find complex and ambiguous relationships in a movement that originally sprang up as a counter to the unfettered forces of industrial capitalism only to find itself taken over in the long-term support of capitalist regimes of accumulation. Nevertheless, it is no coincidence that the transition from an opposing to a supporting role took place at the turn of the twentieth century. Late Victorian society was a transitional society, not only because of the emergence of new social structures and new forms of economic organization associated with industrial capitalism, but also because the growth of scientific knowledge changed the whole outlook of people in relation to their environment.

Planned modernity

Figure 4.1 shows a chronology of examples of key buildings and urban design elements in the twentieth century. In Britain, the catalyst for the emergence of town planning was Patrick Geddes, a professor of biology and a slightly eccentric polymath who, as Peter Hall observes (2002: 143), 'tried to encapsulate the meaning of life on folded scraps of paper'. Geddes was fascinated by cities but appalled by what he saw. He likened them to 'ink-stains and grease-spots' expanding over the 'natural' environment, creating nothing but 'slum, semislum and super-slum' with social environments that 'stunt the mind'. He was an active campaigner for housing reform and had good connections in all the relevant scientific societies and progressive associations. In the early years of the twentieth century, when science was setting discovery after discovery before an increasingly appreciative world, the social sciences in general and the study of urbanization in particular were still in their infancy. In this context, the voice of Patrick Geddes was persuasive. As a scientist, he commanded considerable respect, and was able to speak out on social issues without being dismissed as a liberal do-gooder. Nevertheless, Geddes was a product of his time, intuitively anti-urban. More than anything, he believed that cities needed to be managed, just as a farmer might manage fields of crops or herds of animals. The bad had to be eradicated in order for the good to prosper. There had to be a plan for growth and development, which in turn implied an inventory of present resources.

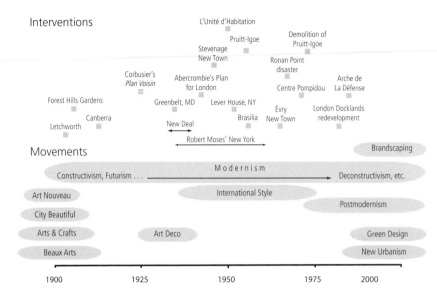

Figure 4.1 Twentieth-century movements and interventions.

It was this idea of an inventory, or survey, that was Geddes's first major contribution to his amateur interest. He had his own model for undertaking such surveys in his Outlook Tower in Edinburgh, which contained a *camera obscura* and a collection of photographs of urban life. Influenced by the writings of French sociologist Frederic Le Play, Geddes believed that the information gathered from urban surveys should clarify the availability of resources and detail people's responses to their physical environment. Following the writings of French geographer Vidal de la Blache, Geddes believed that this inventory should be undertaken within the context of a city's regional framework. This idea was his second major contribution to planning. The region, he argued in his 1915 book *Cities in Evolution*, had to be the basis for the reconstruction of economic, social and political life. Cities and regions needed one another; they had to be planned and managed together, particularly, as he astutely observed, in view of the decentralizing forces of the new 'neotechnic' technologies of electrical power and the internal combustion engine.

Geddes became a leading figure behind Britain's first Housing, Town Planning etc. Act 1909. This was important, in retrospect, not for any significant legislative outcome but for the tenor of the debate that it framed. The Act reverberated with the paternalism, environmental determinism and bourgeois aesthetics that had been slowly developing for fifty years or more and that would continue to shape planning ideology for another hundred. In the words of the Bill introducing the Act:

> The object of the Bill is to provide a domestic condition for the people in which their physical health, their morals, their character and their whole social condition can be improved by what we hope to secure in this Bill. The Bill aims in broad outline at . . . the home healthy, the house, beautiful, the town pleasant, the city dignified and the suburb salubrious.

As John Gold (2007) notes, there was no specification about precisely who would assume responsibility for these ambitious goals. Engineers and surveyors had long exercised statutory responsibilities for tackling basic road planning and sanitation, but architects quickly made a pitch for the new work. Emboldened by rhetoric about the need for a comprehensive view of urban reconstruction

> and lulled into over-confidence by an astonishing naivety about political realities, architects had pressed to place themselves at the centre of the building team. This inevitably brought them into direct competition with the surveyors and engineers, who already had responsibility in this area and saw no particular reason to cede powers to another profession that wanted to increase its own sphere of influence.
>
> (Gold 2007: 67)

Over the next four decades, interrupted by two world wars and the Great Depression, professionalized town planning emerged to assume responsibility. A Royal Charter was granted to the Town Planning Institute in 1914 'to advance the science and art of town planning in all aspects, including local, regional and national planning, for the benefit of the public', but little progress was made by way of legislation or specialized education until after the Second World War.

When progress did occur, it was heavily influenced not only by the nineteenth-century legacy of paternalism, environmental determinism and bourgeois aesthetics but also by the contingencies and developments of the inter-war period. On the one hand, the Depression had generated a vociferous lobby on behalf of impoverished industrial districts and neighbourhoods. On the other, increasing car ownership, combined with unprecedentedly low house prices (in turn a result of inexpensive labour and materials during the Depression), had generated a vociferous lobby of rural conservationists who were alarmed at the rate of urban sprawl. The two groups formed an unlikely alliance in support of stronger planning legislation, with the result that several royal commissions were established, reporting in the midst of the Second World War on Britain's industrial population (the Barlow Report in 1940), rural land use (the Scott Report in 1941) and the issues involved in public control of land use – primarily, increased land values ('betterment') and questions of compensation to private land owners (the Uthwatt Report in 1942). Meanwhile, the Beveridge Report in 1942 laid out the potential contours of a redesigned welfare state.

A landslide Labour victory in the 1945 general election gave the government a mandate to implement a sweeping programme that included the selective nationalization of major industries and utilities (including the Bank of England, coal mining, the steel industry, electricity, gas, telephones, the railways, road haulage and canals) and the establishment of the 'cradle to grave' welfare state conceived by Beveridge. This package of radical reform included two pieces of legislation that formed the basis of urban planning: the New Towns Act, 1946 and the Town and Country Planning Act, 1947. The former allowed the central government to designate local areas as new towns as a partial solution to the pressing problem of rebuilding urban communities after the ravages of the war; the latter required local authorities to develop and implement local land use plans.

Universities quickly stepped in to develop degree and diploma courses to supply the necessary pool of professionals. Their curricula, like the Town Planning Act, 1909, were coloured by paternalism, environmental determinism and bourgeois aesthetics, by Geddes's survey approach, and by the case law and practical experience inherited from the pre-war work of surveyors and engineers. The overall aim of the emergent profession was to tidy up and modernize cities, clearing out the legacy of unregulated urbanization and making them more efficient for business as well as more equitable and amenable for residents. Tactically, enormous emphasis was placed on the separation of land uses as a rational basis for development: keeping like with like, keeping industry away from residential and recreational areas, and keeping suburban development from creeping into farmland. This was an idea that had been first articulated in 1918 by French architect Tony Garnier in his book *Une cité industrielle*. Strategically, the conventional wisdom was derived from Patrick Abercrombie's plan for London in 1943, which rested on a combination of three main elements:

1   Inner-city slum clearance.
2   A designated 'Green Belt' (about 8 kilometres deep) and an 'Outer Country Ring'. Recreation was intended to be the predominant use in the former, and agriculture in the latter. Strict development controls would limit building in this zone and so, it was envisaged, limit sprawl. City dwellers would have access to this green open space 'from doorstep to open country' by way of riverside walks, footpaths, bridle tracks, green lanes, bicycle trails and parkways.
3   A series of new towns, located beyond the Outer Country Ring, to accommodate new industrial spaces and a workforce of 'overspill' population from the thinned-out inner-city slums.

With the implementation of the New Towns Act, 1946 and the Town and Country Planning Act, 1947 the plan became reality, and soon became a showpiece of the idea of 'winning the peace' through planned modernization (see Case Study 4.1).

Case Study 4.1 **New towns for new societies**

In the climate of economic recovery and reconstruction in Europe after the Second World War, there was an opportunity to rethink the design of settlements and a strong sense of optimism that planning and urban design could make a significant contribution to progressive social change. In the United Kingdom, the Barlow Report and Abercrombie's Plan for London provided the rationale for the development of new towns that were to be sited around London and a few other large cities in order to initiate the decentralization and diversification of industry and the reorganization of congested urban areas.

The first generation of fourteen new towns, eight of them around London, were designed very rapidly and built between 1947 and 1950. For each town, the central government appointed a development corporation with its own staff of planners, engineers and administrators with powerful capacity to plan, buy and prepare the land and to build and manage housing, shops, factories and public buildings. They were conceived in direct continuity of planning thought from the utopian visions of Ebenezer Howard and the garden city movement. They were to be located at least 25 miles from London (or 12 miles from other cities), as far as possible on greenfield sites, and were planned for a population of 20,000 to 60,000 people. The density was to be low: just 12 people per acre. A second generation of British new towns was created around 1956. This group included Cumbernauld in Scotland – an experiment in higher densities, traffic separation and a town centre megastructure – and Runcorn, near Liverpool, an experiment in a separate rights-of-way transit system.

A third generation of new towns, more related to regional development policy than to large city overspill and generally not on a completely virgin site, was initiated in the mid-1960s and included Milton Keynes, midway between London and Birmingham. By 1980, 33 British new towns had a total population of 2.4 million and more than 1.1 million jobs. But although reasonably successful in terms of industrial policy and metropolitan deconcentration they were much less successful in terms of design, attracting a great deal of criticism from both design professionals and residents for their rather sterile appearance and placeless identity.

In Scandinavia and the Netherlands, postwar new town development was inspired more by the Modernism of the Athens Charter than by the garden city movement. Amsterdam-West, for example was built in the 1950s as a large new town (for 135,000 inhabitants) to the design of Cornelis Van Eesteren, the general secretary of the Athens Charter group. Following Modernist principles, Amsterdam-West featured a mixture of three- or four-storey apartment buildings (60 per cent), individual terrace housing (25 per cent) and high-rise blocks (15 per cent); a hierarchy of roads with separation of pedestrian and vehicular traffic and small neighbourhood units with their own open spaces, primary schools, corner shops and churches. The quality of the detailing made it a critical success, and the approach was repeated in Buitenveldert, south of Amsterdam, and Nieuwendam, to the north.

In Finland, meanwhile, progressive Modernist principles were used in planning Tapiola, a garden city near Espoo, about 15 kilometres from Helsinki (Figure 4.2). This project was conceived by Heikki von Hertzen and realized through Asuntosäätiö, the non-profit Housing Foundation of Finland. The overall design was executed by Aarne Ervi, who combined modern apartment blocks, terraced houses and detached properties side by side with one another amid 650 acres of forest land. The predominance of apartment buildings provided an impression of low density in a garden-city version of Corbusier's *Ville Contemporaine*. Most of Finland's better-known architects and planners, including Alvar Aalto, Aulis Blomstedt, Jorma Järvi, Raili and Reima Pietilä, Viljo Revell, Osmo Sipari, and Kaija and Heikki Siren, produced plans of a district or of social housing blocks or individual houses, and most of them found their way into architecture journals as examples of good housing design. By the 1960s, Tapiola had become known internationally as an outstanding example of a Finnish Waldstat, 'living next to nature'.

Figure 4.2 Tapiola, Finland. Built in the 1950s and 1960s as a garden city, designed by a group of architects headed by Aarne Ervi. (Photo: Charles & Josette Lenars/Corbis)

In France, the new town programme did not get under way until the mid-1960s, when they were part of a strategic planning response to the super-dominance of the Paris metropolitan region. As a result, French new towns were conceived as part of a planned pattern of regional development. Eight new towns were proposed for the Paris region and five were finally approved in the early 1970s: Cergy-Pontoise in the northwest, Saint-Quentin-en-Yvelines in the southwest, Evry and Melun-Sénart in the southeast, and Marne-la-Vallée in the east. Four other new towns were approved near the provincial cities of Lille (Lille-Est), Rouen (Le Vaudreuil), Lyon (L'Isle d'Abeau) and Marseilles (Rives de L'Etang de Berre).

Etablissements publics (similar to the British new town development corporations) were created in order to plan the new towns and to acquire and prepare land and sell it back to developers of housing or manufacturing. French new towns have no fixed limits and have been developed within existing suburbs as well as in newly developed areas. Their common denominator is a new mixed-use urban centre with residential areas, manufacturing zones, shopping and leisure amenities, and public open space. In Evry, the 'Agora' became the most important multipurpose cultural and leisure centre in France.

Key readings

Anthony Alexander (2009) *Britain's New Towns Past and Future: From Industrial Sprawl to Sustainable Communities.* London: Routledge.
Pierre Merlin (1980) 'The New Town Movement in Europe', *Annals of the American Academy of Political and Social Science*, 451: 76–85.

The American way

In the United States, broadly parallel events, together with the international dissemination of ideas, resulted in a broadly similar story in terms of the emergence of professionalized planning as a response to the perceived need for the planned modernization of cities. But the American emphasis on the sanctity of private property, the supremacy of individual over collective rights, and the dominance of the motor car in economic and cultural life meant that the details were somewhat different. The mobilizing force behind urban planning was the Regional Planning Association of America (RPAA), formed in 1923 as an association of reform-minded individuals that included Clarence Stein (an architect and proponent of the garden city movement), Benton MacKaye (forester and conservationist), Lewis

Mumford (writer and critic), Alexander Bing (a developer and real estate investor) and Henry Wright (an architect and landscape architect). The RPAA's intellectual roots were distinctly radical and owed a lot to Patrick Geddes. It fell to Mumford to translate Geddes's radicalism to the American context, fusing it with the more restorative gospel of the garden city movement and the conservative social engineering ideals of the City Beautiful movement.

Mumford and the RPAA exercised enormous influence on American planning practice and planning ideology from the 1930s onwards, not least through the legacy of Franklin Delano Roosevelt's New Deal programmes. The market failures and exigencies of the Depression undermined the legitimacy of classical, laissez-faire liberalism and led to its eclipse by an egalitarian liberalism that relied upon the state to manage economic development and soften the unwanted side effects of free-market capitalism. An immediate target of New Deal policy was the stimulation of the labour-intensive construction industry by stabilizing the mortgage market and facilitating sound home financing on reasonable terms. This had the effect of reigniting suburban growth, creating a 'spatial fix' to the economic crisis (Checkoway 1980; Harvey 1985). Whereas housing starts had fallen to just over 90,000 in 1933, the number of new homes started in 1937 was 332,000, and in 1941 it was 619,000. Optimistic New Deal administrators saw a further opportunity: to *plan* suburban development, democratizing the suburbs by drawing people from redeveloped central cities. Rexford Guy Tugwell, appointed by Roosevelt as head of the Resettlement Administration, envisaged some 3,000 'greenbelt' towns that would contain government-sponsored low-cost housing, and promptly drew up a list of twenty-five cities on which to start. Funds were allocated for only eight, however, and Congress, under strong pressure from the private development industry, whittled this number down to five. Two of the five were blocked by local legal action. Only the remaining three were built: Greendale, southwest of Milwaukee, Greenhills, near Cincinnati, and Greenbelt, just north of Washington, DC.

Meanwhile, the principle of separating land uses found its way into American urban planning practice through a distinctively American route, with thinly veiled exclusionary undertones. In a landmark case – the *Village of Euclid, Ohio v. Ambler Realty Co.* (1926) – the US Supreme Court established the power of local governments to 'abate a nuisance', ruling in favour of the municipality's right to prevent a property owner from using land for purposes other than for which it had been zoned. A nuisance, the Court ruled, could be defined very broadly to include anything affecting the general welfare of a residential area. As a result, zoning promptly came to be used to exclude not only undesirable land uses from residential areas but also (by establishing large – and therefore expensive – minimum lot or dwelling sizes, for example) undesirable people.

Exclusionary zoning, together with the arrival of mortgage insurance for lenders, the increasing affordability of cars and of the application of production-line manufacturing techniques to large-scale housing subdivisions, laid the foundations for contemporary suburbia. Frank Lloyd Wright, a gifted architect who also fancied himself as a visionary intellectual, could see it coming. Wright, ever the iconoclast, decided to be an architect who hated cities. He was also anxious to position himself as a visionary with a distinctively American flavour. His vision was for a 'Usonian' (a word-play on US own) future. In contrast to the dominant Modernism espoused by fashionable European architects and planners, Wright argued for a low-density, low-rise pattern of settlement in his idealized *Broadacre City*, unveiled in 1932.

Drawing on the individualism and naturalism of Jefferson, Thoreau and Emerson, Wright took a stance that gave primacy to individual freedom rather than to Modernism's emphasis on social democracy. Single-family homes on one-acre lots, he argued, provided the only way to guarantee the individual freedom that was the birthright of Americans; Broadacre City would be the ultimate expression of a truly democratic society. Further, it would be healthful, aesthetically pleasing, and morally and culturally uplifting. The inaccessibility inherent to the large lots and low densities of Broadacre City was to be conquered by a network of land-scaped parkways and freeways, with the focal point of semi-rural neighbourhoods being provided by huge gas stations, architectural centrepieces that would double up as cafeterias and mini-marts. Unfortunately, like most would-be visionary architects, Wright did not really understand cities and their complex interdependencies. Like his predecessors, his contemporaries and his successors, he was able to see no further than a prescriptive and deterministic relationship between urban design and individual and social well-being.

While Wright was dreaming the ideal future American settlement pattern, another key figure was setting about modernizing the existing urban fabric. Robert Moses was a powerful bureaucrat and power broker in New York (both the city and the state) from 1924 to 1968 who led a vast building programme aimed at modernizing urban infrastructure, expanding the public realm with extensive recreational facilities, removing blight, and making the city more liveable for the middle class (see Case Study 4.2). The comparison with Haussmann is unavoidable, but Moses, as Malcolm Miles (2001) observes, oversaw an historically specific kind of urban development in which the American Dream was translated as the freedom to build for money and the freedom to drive. It was also an extreme case of a more general phenomenon: the evangelical zeal of a profession that had been granted the power and resources to reconceptualize the city in a new, modern form.

Case Study 4.2 **Robert Moses: hero or villain?**

Robert Moses began his career as a passionate believer in reform, a vigorous opponent of the favouritism and corruption of Tammany Hall politics in New York. He saw himself as leading a mission to save the city from obsolescence and decline and, like Frank Lloyd Wright, he saw the motor car as the key to the future. Early in his career he developed a reputation as an effective bureaucrat in New York state government but he quickly concluded that he needed to beat the power brokers at their own game in order to get things done. When New Deal tax dollars became available from the Works Progress Administration Program, Moses was one of the few local officials who had 'shovel-ready' projects planned and prepared. Funded by federal dollars, Moses dealt out patronage extensively, building support from labour unions, construction firms, investment banks, insurance companies and real-estate developers. He used his influence to fast-track projects in legislators' home districts; in return, they repaid him by granting money for ever more ambitious projects. His success allowed him, extraordinarily, to hold state and city government jobs simultaneously. At one point, he had twelve separate titles, maintaining four palatial offices across the city and Long Island, and controlling all federal appropriations to New York City. By 1936 Moses was employing 80,000 people on projects that included building a series of parkways in the outer boroughs, developing oceanfront beaches, refurbishing Central Park Zoo, and building 255 playgrounds and 11 outdoor public swimming pools (Schwartz 1998).

His influence increased after the Second World War and he used it to squash the development of a city-wide Comprehensive Zoning Plan that would have restrained his power. By now almost unchallenged, he turned his attention increasingly to 'tower in the park' urban renewal projects and to extending his highway and bridge-building programme, using tolls collected from the new bridges and tunnels to finance ever more projects. Responding to critics who pointed to the displacement of tens of thousands of households and the demolition of historic buildings, his attitude was: 'When you operate in an overbuilt metropolis, you have to hack your way with a meat ax' (Miles 2001: 33). By the time of his retirement in 1968 he had presided over the construction of thirteen road bridges (including the George Washington Bridge and the huge Verrazano-Narrows Bridge), more metropolitan superhighway (670 kilometres) than in Los Angeles, and 28,000 apartment units on urban renewal sites. His achievements were not even-handed, however. His highway projects on Long Island followed a circuitous path so as not to upset wealthy landowners and his entire strategy was based on his idea of creating a more liveable city region for the white middle classes. He actively resisted the use of public transport and steadfastly refused to accommodate plans for subway, bus and train improvements, while critics have pointed out that a pattern of barriers to access for non-white citizens – steep stairs or busy highways, for example – appeared repeatedly in his public projects.

Key reading

Hilary Ballon and Kenneth T. Jackson (eds) (2008) *Robert Moses and the Modern City: The Transformation of New York.* New York: W.W. Norton.

Evangelical bureaucrats

Throughout Western Europe and North America, the postwar period between 1950 and 1980 was characterized by a proliferation of government programmes for housing, urban renewal, land use zoning, transportation planning, environmental quality and comprehensive planning projects. All of these provided jobs for planners and enhanced the profession's visibility and growth. Planners became indispensable links between various kinds of projects and various layers of government and so there was an unprecedented growth in the number of urban planners, in their collective power as a profession, and in their self-confidence concerning the possibility of delivering better, safer, nicer and more efficient cities.

Higher education shifted into gear: the number of planning programmes in the United States, for example, increased from about 20 in the mid-1950s to around 90 by the end of the 1970s. The output of formally qualified planners increased over the same period from about 100 per year to over 1,500 per year. Their educational curricula were informed and inspired by the bold and utopian ideas of Ebenezer Howard, Patrick Geddes, Frederick Law Olmsted, Daniel Burnham, Le Corbusier, Frank Lloyd Wright and Clarence Stein, and by tales of a modern professional history featuring can-do pioneers like Robert Moses. From these roots there developed an evangelical spirit to the entire profession: cities *could* be better places; they *should* be. The influence of European Modernism and the Congrès Internationaux d'Architecture Moderne (1928) and the subsequent publication of its proceedings in the Charter of Athens (1943) added a penchant for rationalistic, sweeping, futuristic solutions (Mumford 2000). Equipped with the latest developments in social science research and theory – the languages and toolkits of behavioural theory, regional economics, regional science, quantitative geography, systems analysis and transportation modelling – the stage was set for a golden age of Modernist planning on a truly heroic scale.

The outcome was perhaps nowhere more visible than in France, where a succession of presidents vied with one another to sponsor *grands projets* in Paris that would remodernize the city a century after Haussmann; and at the same time attest to the stature of the city, of France, and themselves (Fierro 2006). First was the Centre national d'art et de culture (Figure 4.3), pushed through by President Georges Pompidou and renamed the Centre Pompidou after his death in 1974. Not to be outdone, his successor, Giscard d'Estaing, sponsored three *grands projets*: the museum and library complex of the Institut du Monde Arabe; La Villette, a complex at the northeastern edge of the city that includes a vast science museum, a music centre and an exhibition hall set in park studded with massive Modernist sculptures; and the Musée d'Orsay, a new museum housing the Louvre's nineteenth-century collections within a restored railway station.

The projects associated with François Mitterand, d'Estaing's successor, were even greater in scope: the expansion of the Louvre and the addition of a new entrance within a glass pyramid; L'Opera de la Bastille, a new opera house at the Place de la Bastille; the monumental Arche de La Défense (Figure 4.4), the centrepiece of the office complex at La Défense and the culmination of the vista from the Arc de

Figure 4.3 The Centre Pompidou, Paris. Designed by Renzo Piano and Richard Rogers and built in the early 1970s, the building's exposed skeleton of colour coded tubes for mechanical systems made for a radical architectural statement for the city. (Photo: author)

Figure 4.4 The Arche de La Défense, Paris. Completed in 1990, the Grande Arche completed the line of monuments that forms the *Axe historique* running through Paris and was one of the architectural *grands projets* initiated by President François Mitterrand. (Photo: Owen Franken/Corbis)

Triomphe; and the Bibliothèque Nationale (now called the Bibliothèque François Mitterrand), four L-shaped glass towers facing inwards towards each other (symbolizing open books) around a sunken garden. Meanwhile, Paris also acquired all of the routine imprints of modern planning: multiple urban renewal schemes, a motorway ring road system (the Boulevard Périphérique), suburban social housing schemes (banlieues), new towns (Cergy-Pontoise, Marne-la-Vallée, Sénart, Evry and Saint-Quentin-en-Yvelines), controversial modern buildings (notably the Tour Montparnasse, a 210-metre (689-ft) tall office skyscraper of darkly monolithic appearance) and planning *causes célèbres* such as the giant hole in the ground left for years after the demolition of the old food wholesale market of Les Halles, only to be filled by a mediocre shopping centre.

While the *grands projets* in Paris were in many ways the result of a singular politics, cities everywhere were recast through strategic plans that were based on the modernizing principles of strict separation of land uses, slum clearance, large-scale civic and commercial renewal projects, urban expressways and, in Europe, new towns and public housing schemes. In Britain, successive governments increased the targets for new housing construction, with the expectation that one in two dwellings would be in the public sector, where government subsidies encouraged industrialized construction methods and medium- and high-rise buildings. Every large city also had its flagship city-centre redevelopment schemes. As John Gold (2007) observes:

Bold decisions on large-scale clearance and renewal seemed to suggest, even demand, large-scale and imaginative solutions. . . . Psychologically, there was a deep-rooted sense that this was a key moment in the lives of cities. City centres were the heart of urban life. Failure to grasp modernity and introduce change could leave central areas languishing, damage local businesses and condemn a town to second-class status in relation to regional rivals. By contrast, positive and uncompromising decisions could say much about its thrusting, progressive and dynamic nature. Civic pride and place promotion were at stake as well as the need for modernization.

(Gold 2007: 116)

Too often, though, the result was a complex of slab-like buildings and parking decks, often in the Brutalist style of reinforced concrete, with pedestrians segregated from traffic by means of walkways and windswept flights of stairs. And in some cases, as Gold notes, 'it was hard to avoid the underlying sense that civic pride was pushing forward ever-grander schemes that bore little relationship to reality' (Gold 2007: 136).

Although the professional press and academic community were able – through a combination of hubris and evangelical zeal – to keep popular opinion and uncomfortable sociological evidence at bay and maintain their optimism about technology and progress, the golden era of urban planning had come to a close by the mid-1970s. The desirability and effectiveness of planned modernity had been challenged as early as 1961 by Jane Jacobs in her book *The Death and Life of Great American Cities*. Jacobs reasoned that planning had taken away the life and vitality of cities, tearing out their sclerotic hearts only to replace them with a 'great blight of Dullness' in the form of high-rise apartment blocks. Adherence to the dogma of land-use segregation, she pointed out, resulted in the loss of vitality and serendipity in urban life. Left to planners, she argued, city landscapes 'will be spacious, parklike, and uncrowded. They will feature long green vistas. They will be stable and symmetrical and orderly. They will be clean, impressive, and monumental. They will have all the attributes of a well-kept, dignified cemetery' (Jacobs 1958: 157).

Within a decade, public confidence had been sapped by a series of lengthy, costly and ultimately abortive development sagas, and by the whiff of corruption surrounding many of the larger redevelopment schemes. In Britain, the shortcomings of industrialized high-rise housing were dramatically exposed by the Ronan Point disaster in 1968, when the corner of a twenty-three-storey block in Canning Town, London, collapsed after a simple kitchen accident. In the United States, the dynamiting in 1972 of the Pruitt-Igoe project, a group of thirty-three public housing apartment blocks in St Louis (Figure 4.5), provided another critical moment. Seventeen years before, the project had won an award from the American

Figure 4.5 The demolition of the Pruitt-Igoe project in St Louis, 1972. The televised event has been claimed by some as marking the moment that Modern architecture died. (Photo: Bettmann/Corbis)

Institute of Architects. Good as Pruitt-Igoe looked on the drawing board, it turned out to be unliveable: it was the tenants themselves who suggested that their homes be dynamited.

John Gold (2007), writing about the end of the golden era in Britain, writes that by 1972 the architects and planners

> who had previously enjoyed general endorsement as midwives of essential urban transformation . . . now found themselves vilified as dictatorial figures who betrayed public trust and imposed unwarranted change on society – largely by exploiting their eagerly cultivated status as experts. The erstwhile creators of environments fit for tomorrow's society were rebranded as manipulative social engineers. Root-and-branch criticism, not always accurately targeted, had fallen on aspects of road planning, neighbourhood design, land use policy, town centre development and, above all, public sector housing.
>
> (Gold 2007: 12)

For better or worse, planning was no longer any real force for progressive socio-economic change in cities. The dialectics of urban development had brought it entirely, rhetoric aside, into the service of the powerful and the power brokers. The public, its putative beneficiaries, no longer believed or trusted in its abilities or intentions. Almost immediately, the postwar economic boom that had sustained planned modernization came to an abrupt end with an international economic crisis triggered by the sudden quadrupling, in 1973, of the price of crude oil by the Organization of the Petroleum Exporting Countries (OPEC). No longer able to afford grand redevelopment schemes, cities turned to planning departments to assist in branding rather than in engaging real change (see Chapter 5). Planning departments, retreating from their authoritative position and their comprehensive mandate, sought to retain a role instead through public–private partnerships and 'smart growth' strategies whose purpose and direction was almost always dominated by the private sector. Planning professionals turned quietly to local zoning issues, eventually seeking to regain influence and credibility by taking up 'green' issues and by revisiting, selectively, nineteenth-century utopian ideals, joining with architects in suturing artful fragments of neo-traditional developments and mixed-use megastructures to the splintering fabric of increasingly polarized cities (Chapter 5).

Modernism as ideology

All through the era of planned modernity, the dominant design ideology was that of Modernism. It should be acknowledged at once that not only has Modernism found expression across a wide range of art and design, but also it has changed in character and intention over time. Different advocates, innovators and adherents have stressed different aspects of Modernism as a design ideology. The diverse and sometimes conflicting contributions to Modernism have generated an extensive literature, documenting the rich and complex interdependencies among artists, practitioners and the changing political economy of urbanization (Tafuri and Co, 1986, 1991; Frisby 2001; Frampton 2007; for an alternate perspective, see Wolfe 1981). This section outlines the major impulses that have left their mark on cities, and on contemporary thinking about architecture and urban design.

The beginning point has been set by many critics at 1907: the year Picasso painted *Les Demoiselles d'Avignon*, thus initiating the abstractionist challenge to representational art. Art critic Robert Hughes (1980) has suggested that the Cubism of Picasso and Bracque was at least in part an attempt to capture the experience of constantly altering landscapes as seen from a moving train or car. Architects took their cue from this abstractionism, seeking to design buildings appropriate to the needs and experiences of the machine age, translating the angularity of Cubism into built form. Several future-oriented movements had already emerged in response to the challenge of the machine age. In Vienna between 1900 and 1914 there was an extraordinary mix of the creative and the pragmatic in the climate of friction created by the decline of the old imperial era and the beginnings of modernization and scientific inquiry. For a while, Vienna was the pre-eminent centre of art, architecture, music, philosophy, political theory and psychoanalysis: a meeting place of talents, distinctive interests and forceful personalities (Beyerle 2008). Among the radical artists there were Gustav Klimt, Oskar Kokoschka and

Egon Schiele; among the architects and designers were Josef Hoffmann, Adolf Loos and Otto Wagner. Most of them subscribed to the Secessionist movement, established in 1897 in protest against the prevailing conservatism of the Vienna art establishment and its traditionalist orientation. Loos, however, repudiated the iconoclastic and often richly decorated styles of the Secessionists. For him, the principal motivation was the elimination of all 'useless' ornamentation from architecture. He was a prolific writer, contributing numerous op-ed pieces and essays that began to establish a body of theory and criticism of Modernism in architecture.

Meanwhile in Germany the Deutscher Werkbund was founded in 1906 as a loose coalition of artists and craft firms aimed at reforming the relationship between artists and industry. Led by Peter Behrens, the architect and chief designer for the German electrical manufacturer AEG, the guiding principle of the Werkbund was that quantity and quality should complement one another. Behrens saw industrialization as the manifest destiny of the German nation, and his factory designs were for muscular temples to technological power. Other members of the Werkbund (including architects Bruno Taut and Walter Gropius), meanwhile, emphasized the need to overcome the alienating aspects of traditional society by leaving behind all the architectural refinements and symbolism of the old order and replacing them with a no-nonsense style.

Such sentiments were to become central to Modernism as it developed in the 1920s. In general, emphasis was on the importance of redesigning entire cities in the name of technological progress and social democracy. One subset, the Futurists, wanted nothing at all to do with the past. Led by Filippo Marinetti and represented in the field of architecture and urban design by Antonio Sant'Elia, they sought to provoke social and institutional change through cities that were to be stages for permanent revolt, with huge and spectacular edifices that were at once monuments to the masses and to technology. Sant'Elia's drawings of La Città Nuova (1914) showed massive concrete buildings with soaring towers and huge parapets, enormous generating stations and huge factories and airfields, the like of which were not to appear in real urban landscapes for another fifty years.

It was from these roots that there emerged in the 1920s a Modernism that was founded in the idea of architecture and design as agents of social redemption. Through industrialized production, modern materials and functional design, architecture could be produced inexpensively, become available to all, and thus improve the physical, social, moral and aesthetic condition of cities. In France, Le Corbusier conceived his basic formula for the relationship between architecture and cities: both should be a 'machine for living'. Without this kind of radical architecture, he implied, the consequence would be political revolution. His early work involved writing, painting and designing single-family residences (Figure 4.6) and gained him a reputation as a gifted artist and architect. In 1922, he published his ideas on the principles of urban design in a book entitled *La Ville Contemporaine*. The key, he argued, was to reduce the congestion of city centres by increasing their density: by building upwards, in other words. High-density, high-rise city cores, he argued, would leave plenty of space for wide avenues to carry motor traffic and for green space for recreation. La Ville Contemporaine was conceived as a class-segregated city, with the best located, most spacious and best appointed tower blocks reserved for elite cadres of industrialists, scientists and artists. Blue-collar workers were to have smaller garden apartments located in satellite units at some distance from the central cultural and entertainment complex. His plan for Paris, the *Plan Voisin*, reflected this strategy without any concessions to the existing city or to its inhabitants. The *Plan Voisin* called for eighteen 700-foot towers that would have required the demolition of most of historic Paris north of the Seine. Inside the towers, the apartments ('cells', as he called them) were to be uniform, with the same standard furniture. After the power elite failed to support his ideas and the Great Depression took away the ability of industrialists to back him, he revised his ideas on urban design. In *La Ville Radieuse* (1933), he adopted a rather different stance, arguing that everyone should live in giant collective apartment blocks called *unités*, with a minimum of interior space. The perfectly regulated spaces were to be the locus and shaper of new and better patterns of

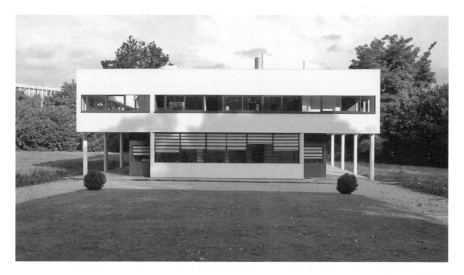

Figure 4.6 The Villa Savoie, Paris. Widely considered to be Le Corbusier's seminal work, the villa in Poissy, outside of Paris, was substantially completed in 1929. (Photo: author)

social life, imbued by a 'new spirit'. His ideas were to become widely influential, but his egomaniacal approach and the failures of many of the urban renewal projects inspired by his work were eventually to give Modernism a bad name (see Case Study 4.3).

Case Study 4.3 **Le Corbusier: hero or villain?**

Within architectural and design circles, Le Corbusier stands alongside Frank Lloyd Wright and Ludwig Mies van der Rohe as the embodiment of modern artistic genius. In broader context, he is recognized as highly influential but his ideas on urban design and planning are widely regarded as suspect or – at worst – nightmarish.

In an artistic and intellectual climate that had become highly politicized, dominated by utopian and futuristic manifestos, Le Corbusier got himself noticed not only for his undoubted design talent but also as a result of his deliberately shocking slogans and outrageous claims for his prescriptions. Le Corbusier would not tolerate any thought of moderation in his approach. Standardization, anonymity and purity of form were essential, he asserted, in creating a 'mass production spirit' that would properly shape the daily routines, desires and leisure activities of the industrial proletariat in modern cities. (Meanwhile, though, he was not above taking commissions for luxury homes from wealthy individuals.)

Le Corbusier claimed not to be able to understand opposition to his prescriptions, choosing instead to regard his own ideas as beyond the grasp of ordinary citizens. But the heroic scale of his ideas and his sheer irrepressibility drew admiration from architects and urban designers who wanted leadership and recognition for their profession. From this there grew a conventional wisdom that cities urgently needed to be modernized through ruthless redevelopment, tearing out their centres and replacing them with high-rise housing linked by urban expressways. Little of this was to be accomplished before the Second World War, leaving Le Corbusier himself to hustle for work from Stalin's USSR, Mussolini's Italy and from the Vichy government during the Nazi occupation of France (Weber 2008).

In the years after the Second World War he transformed architecture again with a renewed focus on individual structures. His work ranged from expressionist designs in sculptural concrete like his famous chapel at Ronchamp in France to cubist-inspired angular concrete buildings. By far the most influential of all his buildings was the Unité d'Habitation (built 1947–1952), a twelve-storey apartment block for 1,600 people in Marseilles (Figure 4.7). It was constructed as a rectilinear ferro-concrete grid into which pre-cast individual apartment units were slotted, like 'bottles into a wine rack' as Le Corbusier put it. With its integrated community services, daycare facilities and shops, it carried through his formula of an apartment building as a 'machine for living'. The building's functional lines not only struck the right chord with the design community but also proved to be relatively inexpensive to construct and amendable to prefabrication, and thus attractive to developers and city governments. It was not long before cities everywhere were being recast in the image of the Unité, often as cheaply as possible and in many cases being imposed in thoughtless and ruthless ways as part of urban renewal schemes. The overall result prompted pre-eminent urban geographer-planner Sir Peter Hall to observe that 'The evil that Le Corbusier did lives after him; the good perhaps interred with his books, which are seldom read for the simple reason that most are unreadable' (Hall 2002: 219).

Figure 4.7 The Unité d'Habitation, Marseilles. Following Le Corbusier's dictum that buildings and cities should be 'machines for living', the Unité d'Habitation incorporates shops, sporting, medical and educational facilities, and a hotel. The flat roof is designed as a communal terrace with sculptural ventilation stacks and a swimming pool. (Photo: Thomas A. Heinz/Corbis)

In Germany, the Neue Sachlichkeit (new objectivity) movement brought together artists and architects who rejected the sentimentality of traditional art and the emotional affect of expressionism. Artists like George Grosz depicted stark inequalities and contradictions of industrialized cities in raw, provocative, and harshly satirical terms. Architects like Erich Mendelsohn, Ernst May and Bruno Taut designed no-frills, functional apartment buildings for working-class families. It was the Staatliches Bauhaus, though, that did more than anything to promote the ideals of the new objectivity. In its fourteen-year history the Bauhaus brought together a remarkable concentration of some of the greatest artists and designers of the twentieth century: Josef Albers, Marcel Breuer, Wassily Kandinsky, Paul Klee, Ludwig Mies van der Rohe, László Moholy-Nagy and Oskar Schlemmer as well as the director, Walter Gropius.

The Bauhaus became the 'refinery of . . . European avant-gardes' (Tafuri and Co 1986: 116), setting the standards for modern design in everything from teapots to workers' housing. The unifying themes were simplicity of line, plain surfaces, suitability for mass production and the notion of the type-form as the optimum solution to the functionality of every product (see Chapter 2). Neither the progressive art nor the socialist ideology that was clearly embodied in the school's approach went down well, however, with German fascism, and the closure of the Bauhaus in 1933 was among the first of the Nazi suppressions after Hitler came to power – an act that contributed to the subsequent canonization of the Bauhaus as the font of Modernism (Volkmann and de Cock 2006).

By that time, though, the ideology of Modernism had begun to spread far and wide. In Russia, the 1917 Revolution had set off a vigorous and sustained debate about the appropriate aesthetic for communal life and production, with various strands of Modernism to the fore, while Modernist graphic art was actively mobilized to meet the communication needs of the Revolution among a mainly illiterate population. It did not take long before heroic and totalitarian ideas about socialist urban design and planning spread to Western Europe, further inciting the prospect of radical prescriptions for European cities. Elsewhere, associations were formed to propagate and discuss Modernism. The most important of these was the Congrès Internationaux d'Architecture Moderne (CIAM), an international association of avant-garde architects formed in 1928. CIAM sustained an intense discourse that led to a manifesto, the Athens Charter in 1943, that set out the ideological basis of modern urban design (Mumford 2000).

In the aftermath of the Second World War, Modernism was widely perceived to be the logical aesthetic for recovery, clean and futuristic. This was particularly so in West Germany, since the Bauhaus became one of the few twentieth-century German cultural achievements that could be assessed positively and used as a basis for developing a new democratic culture. In 1953 Max Bill, a former Bauhaus student, led an attempt to develop a direct successor to the Bauhaus at the Hochschule für Gestaltung in Ulm. It was short lived, however, doomed by interpersonal squabbles and fratricidal disputes over pedagogy. Many of the leading practitioners of the Bauhaus moved to the United States, and Modernism was quickly adopted as the canon in most American architecture schools.

The ground had been set by an exhibition at the Museum of Modern Art in New York City, organized in 1932 by two American architects, Philip Johnson and Henry-Russell Hitchcock. The exhibition included examples of work by Mies van der Rohe and many other European Modernists. More in hope than as a reflection of reality, the exhibition was called the International Style. Gropius subsequently moved across the Atlantic to teach at Harvard, Moholy-Nagy and Mies van der Rohe led the Institute of Design (originally founded as the New Bauhaus) at the Illinois Institute of Technology, and Albers worked at the Black Mountain liberal arts college in North Carolina. Modern architecture and urban design began to appear within North American urban landscapes. In particular, the executives of business corporations were attracted to the visual language of Modernism, seeing it as symbolizing progress and prosperity. In 1952, Gordon Bunshaft of Skidmore, Owings and Merrill showed how Modernism could be deployed in downtown high-rise buildings without having to compromise the form of the building with the stepped set-backs required by land-use zoning (in order to allow light and air to reach the street). His design for Lever House, on New York's Park Avenue (Figure 4.8), simply moved back the base of the steel-and-glass skyscraper, leaving the space left at the foot of the building as a corporate plaza and the building itself as pure form.

A few years later, Mies himself designed an even sleeker steel-and-glass sky-scraper, the Seagram Building, just across Park Avenue from Lever House. Between them, these two buildings immediately set the preferred image for the International Style of corporate skyscrapers. By the 1960s, technological improvements made it possible to have glass facades without having to pay the price of uncomfortably high levels of solar gain. At about the same time, new techniques for mounting glass made it possible to have continuous exterior glass surfaces. Putting the two together resulted in 'glass-box' architecture, buildings that disappear in the reflection of their own surroundings, the ultimate expression of Mies van der Rohe's famous dictum, 'Less is More'. But in the course of its ingestion into the American mainstream, Modernism's commitment to an egalitarian society was edited out, and the social agenda was replaced by one that was almost exclusively commercial. The visual aesthetic remained the same, but its meaning was changing. Modernism now represented capitalism – corporate capitalism – not socialism.

Figure 4.8 Lever House, New York City. Designed by Gordon Bunshaft and completed in 1952, the Lever House, set back from its plaza and with its glass-curtain walls and thin aluminium mullions, was a first realization of the dreams of early Modernists of pure forms of freestanding crystalline shafts. (Photo: Mark Fiennes/ Corbis)

The critique of modern architecture and planning

Without adherence to an idealized social purpose, Modernism became increasingly elitist. In architecture, a distinction emerged in practice between the 'brandscapes' of 'accomodating commercial' work (Klingmann 2007) and the architecture of a self-referential neo-avant-garde, for whom social purpose was replaced by 'critical practice' – which translated into designing projects for a small circle of colleagues, magazine editors and their readership. This was highlighted by the abstract theorizing of iconoclastic 'art compound' architects like Bernard Tschumi, the fantasy sculptural architecture of celebrity architects like Frank Gehry and Zaha Hadid, and the downright elitism of self-selected groups like 'The Whites' (also known as 'The Five': Peter Eisenman, Michael Graves, Charles Gwathmey, John Hejduk and Richard Meier), who approached architecture from the point of view of pure aesthetics, unclouded by social or cultural concerns.

Unhinged from a clear ideological platform and increasingly detached from reality, the profession as a whole held on to the prospect of a techno-utopian future, giving credence to 'visionary' work that proposed city-sized megastructures. Archigram's 'Plug-In City', for example, was a disposable city (with a forty-year lifespan) of prefabricated modular units. Archigram's 'Walking City' consisted of gigantic pods on retractable legs. The pods would hook up to the utility infrastructure of existing cities, then move from one location to another by retracting the legs and setting off, hovercraft style, on a cushion of air.

Not surprisingly, avant-garde architecture in particular and Modernist design in general came increasingly to be regarded as elitist and dysfunctional. Meanwhile, a strong sense of social reform emerged, not from design professionals but from urban social movements, catalysed by the student riots and workers' protests of 1968. The radical climate was soon to be checked by the economic recession that followed the energy crisis triggered by OPEC in 1973, but not before the emergence of a reinforced civil rights movement, fledgling feminist movements, a politicized gay rights movement, and a strong sense of distrust and resentment of big capital and big government. In this new context, the critique of modern architecture and planning flourished. Feminists pointed to the way that the privileging of the visual was linked to a kind of masculinity that involves mastery and detachment. The Olympian god's-eye viewpoint of the city plans, for example, implies a position of power in conceptualizing the city, facilitating the dominance of professionals over that is then reinforced through the opacity of the planning process and the arcana of technical language. Women were 'kept in their place' through comprehensive plans and zoning ordinances that were sometimes hostile, often merely insensitive to women's needs (Bondi 1998; Leslie and Reimer 2003b).

Modern architects, meanwhile, continued to design houses with layouts that reflected and reproduced traditional ideologies of family life and gender relations. Elizabeth Wilson (1991) pointed to the way that modern architecture, self-consciously progressive, had nothing to say about the relations between the sexes. It changed the shape of dwellings without challenging the functions of the domestic unit. Indeed, the Bauhaus helped to reinforce the gender division of labour within households through Breuer's functional modern kitchen. The importance of such designs rests in the way that they present gender differences as 'natural' and thereby universalize and legitimize a particular form of gender differentiation and domestic division of labour.

Others drew attention to the environmental determinism inherent to modern architecture. Oscar Newman (1972) gained a great deal of attention for his attack on Modernist housing. He argued that modern architecture had been too preoccupied with form, with architecture as sculpture, and insufficiently attentive to people's need for functional, defensible spaces. Specifically, he suggested that much of the petty crime, vandalism, mugging and burglary in modern housing projects was related to a weakening of community life and a withdrawal of local social order caused by the inability of residents to identify with, or exert any control over, the space beyond their own front door. Sociability, in short, had been 'designed out' by Modernists, along with colour, variety and ornamentation.

Newman's analysis gave empirical grounding to Jane Jacobs' critical rhetoric; while a series of embarrassments, disasters and scandals (including the Ronan Point disaster and the Pruitt-Igoe debacle among many others) led critics in the popular press increasingly to portray architects as narcissistic, dogmatic, elitist and arrogantly self-regarding, and planners as authoritarian and unaccountable. Tom Wolfe's over-the-top debunking of Modernism and ridicule of its leading practitioners in his book *From Bauhaus to Our House* (1981) became a best-seller, reflecting the popular disrespect for the profession. Alison Ravetz, in her book *Remaking Cities* (1980), described how planning had been transformed from an 'enabling' to a 'disabling' profession as a result of its professional ideology (i.e. its paternalism, spatial determinism, futurism and visual aesthetics) and its evangelical mantle that enabled practitioners to turn a deaf ear to criticism. Drawing on postmodern, post-structuralist, feminist and postcolonial critiques, Leonie Sandercock (1998: 4) characterized planning as having pursued 'anti-democratic, race and gender-blind, and culturally homogenizing practices'.

In short, planners' professional make-up proved tragically unsuited to the ideals that they espoused. Even in their finest hour, they were forced to watch themselves fail. Their evangelism and environmental determinism had led them to get bogged down in 'bureaucratic offensives' like urban renewal and highway construction, to the point where the communities whose lives they had hoped to improve were angry and afraid. Before they knew it, their rationality and their predilection for efficient and tidy land-use patterns had led them to become social gatekeepers.

It was in this climate that postmodern architecture and design began to prosper when the post-1973 recession burst the bubble of postwar growth. The pressures of economic restructuring across fragmented metropolitan space marked the beginning of the end of the golden era of urban design and planning. The idea that cities, along with the economy and society as a whole, could be successfully planned and managed, became increasingly suspect. When economic prosperity returned, it was with a very different cultural sensibility: one of consumer materialism rather than one of radical progressivism, of nostalgia rather than adherence to the promise of a futuristic, Modernist utopia. This is not to deny the very considerable importance of urban design and planning in contemporary metropolitan regions. Rather, as we shall see, they have, for the time being at least, become fragmented and increasingly distanced from any broad sense of social purpose or the public interest.

Summary

Urban design professionals find themselves in a pivotal but problematic position. They are inextricably implicated in the successes and failures of urban development, yet their roles are often ambiguous and ambivalent. They are cast as stewards of the built environment, guardians of the public interest, and champions of the aesthetics of cityscapes; yet at the same time they are servants of their clients and employers. In Britain, professional planning and urban design was influenced not only by the nineteenth-century legacy of paternalism, environmental determinism and bourgeois aesthetics but also by the contingencies and developments of the inter-war period. On the one hand, the Depression had generated a vociferous lobby on behalf of impoverished industrial districts and neighbourhoods. On the other, increasing car ownership, combined with unprecedentedly low house prices had generated a vociferous lobby of rural conservationists who were alarmed at the rate of urban sprawl. The two groups formed an unlikely alliance in support of stronger planning legislation.

In the United States, the emphasis on private property rights, the supremacy of individual over collective rights, and the dominance of the motor car in economic and cultural life meant that the story of planned urban modernization was somewhat different. It was strongly influenced by the ideas of Frank Lloyd Wright, who argued for a low-density, low-rise pattern of settlement in his idealized Broadacre City. Meanwhile, Robert Moses had become a powerful bureaucrat and power broker in New York, leading a vast building programme aimed at modernizing urban infrastructure, expanding the public realm with extensive recreational facilities, removing blight, and making the city more liveable for the middle class. The comparison with Haussmann is unavoidable, an extreme case of the evangelical zeal of a profession that had been granted the power and resources to reconceptualize the city in a new, modern, form.

The quarter-century after the Second World War was the golden age of planned technocratic modernization. In the aftermath of the destruction of the war, Modernism was widely perceived to be the logical aesthetic for recovery, clean and futuristic. This period saw the implementation of the strict separation of land uses, slum clearance, large-scale civic and commercial renewal projects, urban expressways and, in Europe, new towns and public housing schemes. It was all very much in the spirit of the Modernism of the 1920s that was founded on the idea of architecture and design as agents of social redemption. Through industrialized production, modern materials and functional design, it was believed, architecture and urban design could be deployed inexpensively, become available for the benefit of all, and thus improve the physical, social, moral and aesthetic condition of cities. In France, Le Corbusier conceived his basic formula for the relationship between architecture and cities: both should be a 'machine for living'.

But in the course of its incorporation into the mainstream, Modernism's commitment to an egalitarian society was edited out, and the social agenda was replaced by one that was almost exclusively commercial. Without adherence to an idealized social purpose, Modernism became increasingly elitist and detached. At the same time, the implementation of planned modernity often seemed to take the form of 'bureaucratic offensives'. Jane Jacobs, an early critic of urban planning, suggested that the natural vitality of cities was in danger of being 'planned to death'. By the late 1970s, Modernist planning had come under intense criticism as being paternalistic, gender-biased and anti-democratic.

Further reading

Alexander Cuthbert (ed.) (2003) *Designing Cities: Critical Readings in Urban Design.* Oxford: Blackwell. A collection of readings that provide an understanding of the theoretical context from which urban design has emerged, with an emphasis on urban design as a branch of spatial political economy.

Robert Fishman (1987) *Urban Utopias in the Twentieth Century.* Cambridge, MA: MIT Press. An appreciation of the ideas and impact of three of urban planning's greatest visionaries: Ebenezer Howard, Frank Lloyd Wright and Le Corbusier.

John R. Gold (2007) *The Practice of Modernism: Modern Architects and Urban Transformation, 1954–1972.* London: Routledge. Provides a wealth of information on the backgrounds of the architects involved in planned modernity, focusing on their contribution to town centre renewal and social housing.

Jane Jacobs (1993) *The Death and Life of Great American Cities.* New York: Modern Library. First published in 1961, this book was the first to challenge the planned modernization of cities, pointing to the unwanted and unanticipated consequences of urban renewal, zoning and other common practices.

Michael Larice and Elizabeth Macdonald (eds) (2007) *The Urban Design Reader.* London: Routledge. Brings together some of the most influential writing on the historical development and contemporary practice of urban design.

Ali Madanipour (2007) *Designing the City of Reason: Foundations and Frameworks in Urban Design Theory.* London: Routledge. Provides perspectives on how differing belief systems and philosophical approaches impact on city design and development, and explores how this has changed since the impact of Modernism.

Carter Wiseman (1988) *Shaping a Nation: Twentieth-Century American Architecture and its Makers.* New York: W.W. Norton. Examines the major architectural movements of the twentieth century, placing architecture within the philosophical and intellectual thinking of the time.

5 Design for new sensibilities

This chapter examines aspects of architecture, urban design and planning associated with the recent shift in the political economy of urbanization from the egalitarian liberalism of the mid-twentieth century to a 'neoliberal' regime based on free-market doctrines. The role and significance of design itself has changed as urban development has responded to the increasing materialism of popular culture and the increasing entrepreneurialism of city governments. Among the outcomes featured in this chapter are the gentrified neighbourhoods of the inner city; the themed and packaged subdivisions, Traditional Neighbourhood Developments and New Urbanism of residential districts; the rebranding, redevelopment and regeneration schemes of older industrial districts; and the general influence of postmodern design and the 'brandscapes' of consumerist society.

Changes in the political economy of urbanization since the mid-1970s have brought significant adjustments to the content and direction of urban design and planning. Meanwhile, new movements in art and architecture have inscribed new aesthetic elements into both residential and commercial townscapes; and the increasing tendency for new class fractions and affective 'neotribal' groupings to establish their distinctiveness through individualized patterns of consumption has ratcheted up the role of design in every sphere of life. In short, contemporary cities, mostly a product of the political economy of the manufacturing era, have been thoroughly remade in the image of consumer society. Design professionals have had to adapt to a neoliberal political economy in which progressive notions of the public interest and civil society have been all but set aside. Producers, for their part, have developed new product lines in response to changing technologies, building systems, regulatory environments, financial systems and consumer demand. Consumers have developed new preferences and priorities in response to dramatically changing physical environments, social structures and patterns of disposable income. In the

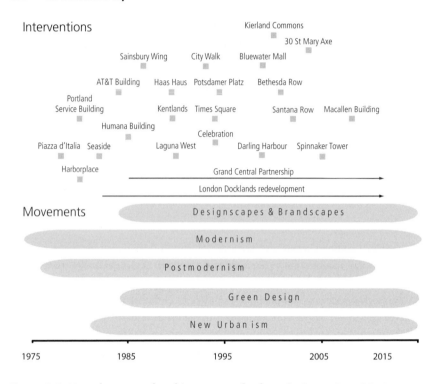

Figure 5.1 Key elements of architecture and urban design, 1975–2010.

process, cities have been reconceived and re-formed. Figure 5.1 shows the timeline of key movements and interventions mentioned in this chapter.

The role and significance of design itself has changed. In the formative years of consumerism, Fordism was the dominant paradigm: mass production for mass consumption. Questions of taste, sophistication and style – mediated by design – were largely precluded by the standardization of mass-produced items whose most important attribute was affordability. Consumers were more concerned with value for money and avoiding below-par experiences than with experiencing the unique or the extraordinary.

But the more that Fordism succeeded in meeting consumer demand for material goods, the more that style and design became important. In the 1980s, 'horizontal desire' (coveting a neighbour's goods) was displaced by a 'vertical desire' (coveting the goods of the affluent, as portrayed in lifestyle magazines and on television). Competitive spending, bolstered by tax breaks, trends in financial markets, the escalating paychecks of the 'new economy' and, not least, by easily available credit, intensified the importance and significance of design. In European and Japanese cities, where consumer durables had to be designed to fit smaller spaces, there emerged an especially strong emphasis on product design.

Over this same period, globalization has rendered cities increasingly interdependent, introduced an increasing degree of cosmopolitanism to urban populations, disseminated the dominant sensibilities of corporate transnational capitalism, and generated complex commodity chains that have resituated the ecology of design services in particular places. Together with structural economic change, globalization has also intensified competition among cities in attracting investment. As a result, design is an increasingly important element of competitiveness for cities in a postindustrial economy. Indeed, as Aspa Gospodini (2002) observes, the traditional relationship between urban economy and urban design seems to have been reversed:

> While for centuries the quality of the urban environment has been an outcome of economic growth of cities, nowadays the quality of urban space has become a prerequisite for the economic development of cities; and urban design has undertaken an enhanced new role as a means of economic development.
>
> (Gospodini 2002: 60)

Place marketing has become central to urban design and planning, while branding – of corporations, products, services and experiences as well as places themselves – has transformed the experiential settings of cities everywhere.

Neoliberalism: new roles for the design professions

The underlying shift in the political economy of urbanization has been one from the egalitarian liberalism of the mid-twentieth century, with its Keynesian economic policy framework and welfare states, to a 'neoliberal' regime based on free-market doctrines. The failure of Keynesianism (the operational policy framework for egalitarian liberalism) to cope with the economic system-shock of the sudden quadrupling of crude oil prices by the Organization of the Petroleum Exporting Countries in 1973, together with the consequent economic recession and the subsequent globalization of industrial production, opened the way for the radically different policy perspectives of neoliberalism.

Affluent households, caught up in an ever-escalating material culture and wanting more disposable income for their own private consumption, were increasingly resentful of taxation. But intensifying materialism and competitive consumption elevated middle-class expectations more rapidly than productivity, with the result that both household savings and public-sector spending were 'crowded out' by private consumption. Consumer appetites were fuelled by increasing credit-card debt; by the deregulation and securitization of capital markets and the consequent expansion of derivatives, hedge funds, collateralized debt obligations and mortgage-backed securities; and by a banking industry that became increasingly competitive and increasingly lenient, offering all sorts of packages (interest-only mortgages, graduated-payment mortgages, growing-equity mortgages, jumbo and super-jumbo mortgages, shared-appreciation mortgages, step-rate mortgages, as well as sub-prime mortgages) to make big mortgage repayments easier for buyers to contemplate. It did not take long for suburban landscapes to reflect the result: a proliferation of monster homes, McMansions, 'starter castles' and the associated 'suburban bling' of 'vulgaria' and 'privatopia' (McKenzie 1994; Knox 2008).

With pressure on public spending, the quality of public services and physical infrastructures inevitably deteriorated, which in turn added even more reasons for those with money to want to spend it privately and to resent paying for public services that they felt they no longer needed. The concept of the public good was undermined as government (to paraphrase Ronald Reagan) came to be identified as the problem rather than the solution. Whereas market failures of the 1930s had been the rationale for the ascendance of egalitarian liberalism, government failures of the 1960s and 1970s became the rationale for neoliberalism. Globalization also played a part: Keynesian economic regulation and redistributive programmes came to be seen as a hindrance to international competitiveness.

Labour-market 'flexibility' became the new conventional wisdom. Neoliberalism regards policies designed to redistribute resources to disadvantaged areas or individuals as necessitating excessive taxation of the wealthy, thereby discouraging entrepreneurial leadership, reducing investment capital and undermining productivity. If necessary, social goals and regulatory standards have to be sacrificed, it is argued, to ensure that business has the maximum latitude for profitability. The rising tide of economic development, the argument goes, will then float all boats, urban and rural, central city and suburban. Thus, rather than focus on metropolitan problems such as poverty, environmental degradation and traffic congestion, urban governance became more concerned with providing a 'good business climate' that might attract investment (Hackworth 2007). As geographer David Harvey (1989a) noted, urban governance had shifted decisively from management to entrepreneurialism.

These changes are part of a very broad shift in governance. Jamie Peck and Adam Tickell (2002) have characterized the process in terms of a combination of 'roll-back' neoliberalization and 'roll-out' neoliberalization. Roll-backs have meant the deregulation of finance and industry, the demise of public housing programmes, the privatization of public space, cutbacks in redistributive welfare programmes, the shedding of many of the traditional roles of central and local governments as mediators and regulators, curbs on the power and influence of labour unions and government agencies, and a reduction of investment in the physical infrastructure of roads, bridges and public utilities. Roll-out neoliberalization has meant the establishment of public–private partnerships, the encouragement of inner-city gentrification, the creation of free-trade zones, enterprise zones and other deregulated spaces, the assertion of the principle of 'highest and best use' for land-use planning decisions, and the privatization of government services.

The overall effect has been to 'hollow out' the capacity of central government while forcing municipal governments to become increasingly entrepreneurial in pursuit of jobs and revenues; increasingly pro-business in terms of their expenditures; and increasingly oriented to the kind of planning that keeps property values high. Neil Brenner and Nik Theodore (2002: 21) suggested that the implicit goal of neoliberalization at the metropolitan scale has been 'to mobilize city space as an arena both for market-oriented economic growth and for elite consumption practices'. As a result, planning practice has become estranged from theory and divorced from any broad sense of the public interest. Planning and urban design have become pragmatically tuned to economic and political constraints rather than being committed to change through progressive visions. Public–private partnerships have become the standard vehicle for achieving change, replacing the strategic role of planning with piecemeal dealmaking. Planning has become increasingly geared to the needs of producers and the wants of consumers and less concerned with overarching notions of rationality or criteria of public good.

A starkly neoliberal political economy, in which progressive notions of the public interest and civil society have been eclipsed by the bottom line in corporate and public–private investment, means that design solutions have to be commercially attractive. Compromises have to be made, projects have to be hustled, ideas have to be sold, and unpalatable truths have to be spun into palatable propaganda. Planning and urban design, deflected from issues involving regulation or social expenditure, were increasingly pushed toward contributing to artful fragments of upscale suburbia as a means of sustaining professional identity and credibility (Knox and Schweitzer 2010). Instead of developing and implementing strategic plans, planning departments were reduced to rubber-stamping the subdivisions and mixed-use 'town centres' proposed by property developers who were surfing the credit boom. All planners could reasonably hope for was to make developments more artful (through themed and packaged design) or at least less artless (through 'smart' growth). Amid the surge of materialism, meanwhile, design in general became increasingly important in the value chain, leaving its mark across every aspect of urban life. Product design, graphic design and fashion – especially prêt-à-porter fashion – simply flourished, while postmodern design emerged as an expression of the hedonistic sensibilities of the affluent and creditworthy classes.

With neoliberalism established as an ideological 'commonsense', it was a short step to what Neil Smith (1996) has called revanchism, and what Sharon Zukin (1991) has memorably described as 'pacification by cappuccino': reclaiming urban spaces from low-income communities and low-profit settings through urban policy and planning in the cause of a 'good business climate'. Cindi Katz (1998), in her discussion of the public–private Grand Central Partnership in New York City, provides an example of some of the ways that such an agenda is pursued, and shows how particular social actors and their activities are removed from view in the interests of ensuring an 'orderly', 'clean' and 'safe' public space. The project, she notes:

> resulted in the removal of all kinds of people from Grand Central and its environs, suggesting that they have, at best, unequal rights to the city. Many of these people, among them the shoeshiners and retailers, earned a modest living in the station and caused no harm to others, but their presence did not seem to fit the new image for Grand Central which includes Michael Jordan's expensive steak house taking up a quarter of the mezzanine, a Godiva chocolate shop, and the redundant but inevitable Starbucks coffee stall. Their lot, like so many aspiring middle and working class people in contemporary New York seems of no moment to the architects of the neo-liberal city – witness the enduring assaults on the City University, the Giuliani administration's 1998 harassment of taxi drivers and restrictions on street vendors, and the searingly high rents for even the smallest commercial spaces. Yet, the texture of the city – its very driving force and unique quality – will be lost if such groups of people have no place in it. If Grand Central becomes as much of an ordered contrivance as a Disney production and its commercial attractions no different from any upscale mall, the Partnership's 'operation' may be considered a success but the patient will be dead.
>
> (Katz 1998: 42)

The proponents of neoliberal policies have advocated free markets as the ideal condition not only for economic development, but also for political and social life. Ideal for some, maybe. Free markets have generated uneven relationships among places and regions, the inevitable result being an intensification of economic inequality at every scale, from the neighbourhood to the nation state. Under neoliberal regimes in the United States and Europe, schools deteriorated while prison construction forged ahead. Public–private partnerships such as the Grand Central Partnership described by Katz (1998) lovingly and lavishly restored, refurbished, preserved and built only those landscapes visible to and valued by the affluent:

> The sanitized visible environment creates a sense of well-being and civic pride for those who count, while public housing decays or goes unbuilt, schools and schoolyards get more crowded and dilapidated, long tended community gardens

in gentrifying neighborhoods are 'condemned' (confiscated) for luxury housing development, and parks and playgrounds in poor neighborhoods are allowed to languish, broken-down and unsafe.

(Katz 1998: 44)

The global financial meltdown of 2008–2009 – triggered by massive losses in the sub-prime mortgage market and compounded by the complex interdependence of global financial markets that had been on a long, speculative and unregulated credit binge – brought not only a sharp halt to consumer spending but also a significant challenge to the underlying cause of the problem: neoliberal ideology. At the time of writing it is too early to tell whether, in the long term, neoliberalism will wane or whether the materialism and 'self-illusory hedonism' of 'romantic capitalism' (see p. 8) will be curbed or perhaps displaced by a new sensibility. Meanwhile, the legacy of neoliberal political economies remains very much in evidence in the form of privatized public spaces, gentrified neighbourhoods and private master-planned communities.

The privatization of public spaces

A series of property booms since 1980 has added significant new spaces to cities. In the initial period after the mid-1970s recession, with poor prospects for profit elsewhere, a great deal of capital investment found its way into the built environment. Much of this investment was in central-city office development, with an eye to the future expansion of service industries. The short-term result was an oversupply of office space in many cities. As economic recovery took hold, developers began to identify consumption – and the experience of consumption – as a key sector. In North America, suburban shopping malls and downtown galleries proliferated; and the more there were, the bigger and more spectacular the next ones needed to be (Knox 1991). In Europe, too, suburban shopping malls and retail complexes appeared, though somewhat constrained in comparison with North America because of the higher cost of land and the lower levels of car ownership.

On both sides of the Atlantic, the competitiveness of property development, combined with the increasing entrepreneurialism of city governments and the increasing materialism of popular culture, meant that development schemes took place at ever-larger scales, with mixed-use complexes and waterfront redevelopments, often built as public–private partnerships, offering integrated settings for mutually supporting, revenue- and tax-generating packages of retailing, offices, residences, hotels and entertainment functions. One of the seminal developments in this context was Harborplace, in Baltimore's Inner Harbor, opened in 1980 as the first part of an extensive makeover of the harbour area. Similar waterfront developments have subsequently been undertaken elsewhere, including South Street Seaport, New York City (Figure 5.2); Fisherman's Wharf, San Francisco; Granville Island, Vancouver; Harbourside, Bristol; Leith's dock area; Dundee Waterfront; Queen's Walk, London; Baltic Gallery, The Sage and the Millennium Bridge, Gateshead; and Darling Harbour, Sydney.

Consumption, scale, spectacle and themed simulations are the common denominators of these developments, as with more recent packages of real estate development in the form of 'lifestyle centres' such as Santana Row (San José, California), CityPlace (West Palm Beach, Florida: Figure 5.3), Bethesda Row (Bethesda, Maryland) and Kierland Commons (Scottsdale, Arizona). These are 'niche' retail concepts driven by the booming consumer goods market of the early 2000s, smaller than traditional malls and without anchor department stores. Rather, they mimic old-fashioned Main Street settings, with tree-lined pavements and

Figure 5.2 Pier 17 at South Street Seaport, New York City. The festival marketplace was built around restored early-nineteenth-century commercial buildings, including the former Fulton Fish Market, and incorporates the National Maritime Museum as part of its tourist attractions. (Photo: Gail Mooney/Corbis)

Figure 5.3 CityPlace, West Palm Beach, Florida. Designed to resemble an Italian town centre, CityPlace is an imagineered 'visitor destination' with over sixty shops, twenty-five restaurants and live entertainment on the plaza.
(Photo: Alan Schein Photography/Corbis)

lampposts, manicured shrubbery, made-up street names and plenty of free parking. Besides containing ubiquitous retail chains like Armani Exchange, Banana Republic, Body Shop, Boss, Crabtree and Evelyn, Diesel, Footlocker, French Connection, H&M, Lacoste, Mango, Samsonite, Sony, Swarovski, Tommy Hilfiger, Urban Outfitters and Zara, they also offer cinemas, upscale restaurants, fitness clubs, outdoor cafés and street entertainment. According to the International Council of Shopping Centers there were just 30 lifestyle centres in the United States in 2002 but by 2007 there were more than 160. Santana Row contains 95 specialty shops and restaurants, with a farmers' market on Sundays. The development stretches 1,500 feet – equivalent to three or four city blocks – an ideal distance for a nice stroll but not too long a walk. Above the stores are the wrought-iron balconies of the housing units on the third floor. With plazas, fountains and street furniture, and outdoor heat lamps to ward off the chill for people who want to linger late into the evening, lifestyle centres are conducive to casual browsing and people-watching.

The urban affect designed into these developments – with their plazas, 'streets', atria, food courts, sculptures, gardens and parking decks – has blurred the traditional distinctions between public and private spaces (Low and Smith 2005). There is a presumption that the spaces that are external to shops, offices, cinemas, and so on are part of the public realm, just like the streets, pavements and parks of traditional urban form. But in reality they are private, and the public is welcome only as long as they do not spoil or threaten the ambience of the spaces as imagineered by their developers, owners and managers. In practice, this means that a good deal of normal urban activity and comportment is edited out of these settings. Peggy Kohn introduces her book *Brave New Neighborhoods: The Privatization of Public Space* (2004) with the example of Stephen Downs, a lawyer who was arrested for trespassing in a mall near Albany, New York, in 2003. 'He did not sneak into the mall after hours or enter some "employees only" part of the property. He was arrested for wearing a T-shirt that he purchased at the mall with the slogan "Give Peace a Chance".' It was on the eve of the war with Iraq, and the message evidently did not suit the desired ambience of the mall.

All sorts of other activities that typically occur in genuinely public spaces – the distribution of leaflets, union picketing, demonstrations and rallies by activists, the solicitation for funds or signatures, the sale of home-baked cookies, and busking, for example – are likewise proscribed from new spaces of consumption. Desert Ridge, for example, a 'town centre' development on the edge of Phoenix, Arizona, has a rigorous code of conduct posted beneath its store directory. The list of forbidden activities includes 'non-commercial expressive activity', 'excessive staring' and 'taking photos, video or audio recording of any store, product, employee, customer or officer' (Blum 2005). It goes without saying that the mere presence of certain categories of individuals – the colourfully eccentric, the boisterously loud, the poor, knots of youths, shifty-looking individuals, the homeless, and so on – is also frequently seen as undesirable by the owners and operators of quasi-public spaces. Though they may not be formally proscribed, these unwanted groups will be deterred through clever design features (like the convex 'bum-proof' benches described by Mike Davis (2006) in Los Angeles) and, if necessary, moved on by private security personnel. Ironically, therefore, just as many of these new spaces of consumption are increasingly designed to recreate the atmosphere of old-fashioned downtowns, they effectively restrict the civic activity, diversity and serendipity that gave mid-century city centres their atmosphere and vitality.

Gentrification

Another characteristic feature of urban change during the neoliberal era has been the gentrification of selected inner-city neighbourhoods. Gentrification involves the renovation of housing in older, centrally located lower-income neighbourhoods through an influx of more affluent households seeking the character and convenience of less-expensive but well-located residences. Typically, the colonizing households are dominated by young professionals involved in the 'new economy', together with teachers, lawyers, designers, artists, architects, writers and creative staff in advertising firms.

These incomers often use 'sweat equity' – their own, do-it-yourself labour, rather than contracted labour – for renovations and improvements. They are, in Bourdieu's terms, typically rich in cultural capital but limited in economic capital. Their arrival displaces poorer households through evictions, escalating rents and house prices, increasing property taxes, and prompting the closure of stores specializing in inexpensive goods and produce. Incoming gentrifiers, meanwhile, contribute to the physical renovation or rehabilitation of the older and usually rather deteriorated housing stock while supporting new businesses such as upscale restaurants, coffee shops, delicatessens, wine bars, galleries, clothing boutiques and bookstores.

The resulting new social ecology, as Sharon Zukin (1991) has observed, functions as a 'critical infrastructure' of cultural change: the habits, tastes and aesthetics of gentrifiers helping to define acceptable codes of conduct, 'good taste' and urban fashionability that are commodified via Sunday newspaper supplements, lifestyle magazines and television programmes on food and home improvements. The fashionability and buzz of gentrifying neighbourhoods, meanwhile, adds to their attraction, drawing in more (and more affluent) households, and more capital investment in businesses. The transitional process is described neatly by Gary Bridge (2006):

> As the gentrification process consolidates, higher-paid professionals are attracted. Sweat equity gentrifiers are replaced by small builders who turn over the properties in the gentrified style. When the neighbourhood is fully 'established', large developers might get involved, involving a routinisation of the aesthetic aspects of the process. Cultural capital in the form of the gentrification aesthetic gets absorbed into the overall 'price' of the property and the neighbourhood in which it is located.
>
> (Bridge 2006: 723)

Because it brings about improvements to the built environment, encourages new retail activity and results in the expansion of the local tax base without necessarily drawing heavily on public funds, gentrification has become an important symbol and prospect for urban change for ideological neoliberals. Because it fosters capital accumulation, caters to the consumption patterns of higher-income groups, and results in the displacement of vulnerable and disadvantaged households, it has become emblematic of urban restructuring and a portent of urban change for liberals. Academic debate about gentrification has been framed around varying interpretations and emphases (Lees *et al.* 2008). The first emphasizes the importance of occupational, social and cultural shifts in influencing patterns of demand. The increased pool of professional, administrative, managerial and technical workers in the new economy, together with the intensification and differentiation of consumption in Western culture, has generated an expanding group of potential gentrifiers. Because many of these potential gentrifiers are employed in city-centre settings, and because their aesthetic sensibilities lead them to reject modern homes in downtown apartments or suburban subdivisions in favour of settings with some history, human scale, and ethnic and architectural diversity, they are attracted to older inner-city neighbourhoods (Ley 1996, 2003).

A second interpretation, associated mainly with the work of Neil Smith (1996), sees the chief actors as real estate agents and developers who are responding to the 'rent gaps' that have resulted from the 'devalorization' of inner-city neighbourhoods. The potential profit from property in older neighbourhoods is often much greater than the profit to be derived from existing buildings and businesses, whose run-down condition may also suppress the profitability of nearby enterprises. This situation is exploited by three types of developers: professional developers, who purchase property, redevelop it and resell for profit; occupier-developers, who buy and remodel property and inhabit it after completion; and landlord developers, who rent to tenants after rehabilitating property. Their efforts are supported by the neoliberal 'roll-out' initiatives of city governments, including business improvement districts, public–private partnerships for mixed-use developments, waterfront redevelopments and festival settings; and the recycling of old industrial and commercial buildings by turning them into concert halls, art galleries, theatres, and retail spaces – what David Harvey (1989b) has dubbed the 'Carnival Mask' of urbanized capitalism.

Writers such as Sharon Zukin (1998) and Rosalyn Deutsche (1988), meanwhile, have emphasized the role of the avant-garde and the contextual shift toward a 'society of the spectacle' (see Chapter 1) in which stylish materialism is increasingly important. Politicians, speculators and developers have come to see edgy art and culture as a crucial element in any new project. Artists and designers themselves, many of whom have a lifestyle that focuses very much on identity, appearance, presentation of self, fashion design, decor and symbolism, have been critically important not only in pioneering the 'resettlement' of rundown, low-rent areas but also in simultaneously providing these areas with the 'designer' touch of radical chic necessary for them to be seen in a new light by the newly affluent and aestheticized professional service workers. David Ley (2003) provides the following quote from an artist in Vancouver, Canada:

> Artists need authentic locations. You know artists hate the suburbs. They're too confining. Every artist is an anthropologist, unveiling culture. It helps to get some distance on that culture in an environment that does not share all of its presuppositions, an old area, socially diverse, including poverty groups.
>
> (Ley 2003: 2534)

On a more prosaic level, many artists are also drawn to 'authentic locations' by the lower rents in neighbourhoods in the early stages of gentrification. As Richard Lloyd (2005) notes, the early arrival of artists in gentrifying neighbourhoods provides a neo-bohemian atmosphere that is attractive to more affluent, educated cosmopolitans employed in the postindustrial growth sectors of finance, insurance, real estate and media technology.

Other influences on gentrification include class, gender and sexuality. Liz Bondi (1991) and Alan Warde (1991), for example, point out that the location of dual-earner households in the inner city is a solution to problems of access to work and home and of combining paid and unpaid labour for married middle-income women and men in well-paid career jobs. This development, of course, is related to other broad socio-demographic trends, including the restructuring of family life that has been reflected in the postponement of child-bearing, decreased family size and closer spacing of children. These changes result in small, affluent households prepared to pay high prices for sought-after housing because they benefit more from the reduction in commuting costs associated with inner-urban residential locations than do those with only one adult working in the city centre.

Conservative utopias: déjà-vu urbanism

A very different market segment emerged from the social polarization that resulted from the neoliberal era and its 'Dream Economy' (see Chapter 1) of intensified consumerism. It is characterized by households that market researchers have labelled 'Innovators' and 'Achievers' – households with high levels of economic capital, but rather less by way of social capital and often less still by way of cultural capital (Knox 2008). These market segments are distinctive because of their conservatism and the emphasis that they place on status, structure, stability and predictability. As consumers, they favour homes, neighbourhoods, products and services that demonstrate their success to their peers. They tend to want houses that make a clear statement about themselves and their lifestyles (basically: 'I've got a big/bigger/better equipped/more spectacular/more luxurious one').

For these resource-rich, aspirational consumers, their house must also be a show-case for the right 'stuff': the furnishings, possessions and equipment necessary for the enactment of their preferred lifestyle and self-image. Developers have responded by producing private, master-planned subdivisions laid out with their own packages of amenities. The result of carefully researched niche marketing and product differentiation, they offer combinations of amenities and themed settings that are matched to the finances and aspirations of different income and lifestyle groups. Some are packaged to appeal to young families, with amenities like water parks, cycle paths, hiking trails, tennis courts, rock-climbing walls, parks, supermarkets, churches, elementary schools and country clubs. Others are packaged to appeal to affluent retirees, furnished with Starbucks cafés, internet access, multi-gyms, tennis courts, pools and golf courses. Some are packaged as 'green' communities or 'sustainable' developments.

The common denominator is the commodification of community and sense of place, carefully cultivated by developers through the theming and branding of their developments, the names that they give them, and the advertising copy they deploy. Gates – real or implied – and perimeter walls and fences are another common feature. They are mainly for show, signalling status and distinction, even though security is often cited as the motivating force behind the trends in gated communities (especially when they are 'forted up' with security patrols and surveillance).

Perhaps the best known and most bizarre of themed and packaged developments is the Disney Corporation's town of Celebration, Florida (Figure 5.4). Founded in 1994 by the Disney Corporation on 10,000 acres of company land just off one of the exit ramps to US Highway 192 about 20 miles south of the city of Orlando, Celebration is a subdivision in neotraditional style, laid out in conscious imitation of towns like nearby Kissimmee, a small Florida mainstreet-style town. It has an imposing town hall but no town government – the town manager is a Disney executive. Nearby is Celebration's school, run by Osceola County but with a curriculum designed and approved by Disney. The homes on Celebration's residential streets are a mixture of classic traditional Southern heritage styles, most of them featuring big front porches and white picket fences. Prospective buyers are asked to sign a 75-page Declaration of Covenants before being permitted to move into their home. Such a document is normal practice in private master-planned subdivisions. In legal terminology it constitutes a 'servitude regime': a set of covenants, controls and restrictions (CCRs) that circumscribe people's behaviour and their ability to modify their homes and yards. Servitude regimes are typically drafted by developers but implemented by homeowner associations, membership of which is mandatory for every home owner in the development. Retailing in the town centre is heavily oriented to tourism, so that everyday shopping requires

Figure 5.4 Main Street, Celebration, Florida. The 'downtown' of the subdivision offers three boutiques, two gift shops, a gallery, two restaurants and a wine bar. (Photo: Preston Mack/Getty Images)

residents to get in their cars and drive to the nearest mall on US 192. As Alex Marshall (2003) observes:

> Celebration is part of the ecosystem of US 192. Its residents depend on the highway's stores to exist. Its businesses have links to wholesalers. Their sewage and water are part of the same system that links those businesses on US 192. Like the shopping centres, hotels, and other subdivisions that line US 192, Celebration is one more pod off of it. . . . The irony here is that Celebration pretends to be the antithesis of US 192.
>
> (Marshall 2003: 233)

The design and marketing of these packaged landscapes of 're-enchanted suburbia' (Knox 2008) has been strongly influenced by discourse in architecture and planning about limiting sprawl, fostering community, civility and sense of place in compact, mixed-use, walkable and relatively self-contained developments. This has drawn on the legacy of ideas and impulses that go back to the intellectuals' utopias of the nineteenth century, the City Beautiful, Geddes's natural region and urban-rural transect; Clarence Perry's neighbourhood unit idea and the precedents of the garden suburbs of the late nineteenth century. Some, like Kirchsteigfeld, in Germany, have been relatively successful in translating these ideas into contemporary context (see Case Study 5.1). But in the compromised professionalism of a neoliberal political economy, it has often found expression in highly commodified and regressive, rather than progressive, form. This can be traced to Traditional Neighbourhood Development (TND), an attempt to codify tract development in such a way as to create the look and feel of small-town, pre-Second World War settings in which pedestrian movement and social interaction are privileged over vehicle use. This is a conservative reaction to social change that mediates cultural and economic anxieties through nostalgic representations of (imagined) better times (Veninga 2004).

In terms of design, Traditional Neighbourhood Developments generally avoid having culs-de-sac because they are thought to inhibit social contact. Front driveways and garages are avoided because they are felt to be ugly, to affirm the primacy of the motor car and to be out of keeping with traditional and vernacular house types. In a similarly motivated attempt to provide guidelines for a new typology of suburban development, architect Peter Calthorpe (1993) developed the concept of the 'Pedestrian Pocket'. Harking back to the days of streetcar suburbs, Calthorpe's idea was for higher-density suburbs to be situated within a quarter-mile walking distance of public transportation hubs: ideally, light rail stations. Thus, pedestrian pockets would become part of a regional scheme of 'Transit-Oriented Development' (TOD). Fine in theory, but the capital costs of new transit lines have relegated TOD, for the most part, to 'paper architecture' – at least until fuel prices rise and car-dependent suburbs become obsolete. Calthorpe's best known plan was for Laguna West, near Sacramento, California. Begun in 1990, the plan called for a walkable development of 2,300 units with a commercial centre in traditional style, eventually to be linked to Sacramento via rapid transit. Marketing and financial pressures, however, meant that the development was modified as it was implemented, resulting in a subdivision that is more scenic and at a higher density than conventional developments, but otherwise similar to other California suburbs, with residents living car-dependent lives.

Traditional Neighbourhood Development, on the other hand, was quickly taken up by developers of private master-planned communities. It did not take long for Modernist design critics to condemn such settings as mawkish, camp, costume drama; and as 'hyperreal' environments based on cultural reductiveness (Eco 1986). Nan Ellin (2006: 99) refers to the nostalgic reflex to 'drag and drop forms from other places and other times' as 'form following fiction'. Architecture theorist Kenneth Frampton (1983), in an early reaction to the glibness of the nostalgic reflex in architecture, called for a more 'critical regionalism' that might assimilate genuine local materials, crafts, topographies and climate with the broader trends

Case Study 5.1 **Kirchsteigfeld, Germany**

Kirchsteigfeld is an entirely new development, explicitly modelled on the traditional morphology of small towns of central Europe. Located on the edge of Potsdam, just 36 kilometres from Berlin, Kirchsteigfeld has a population of around 6,000 on a compact site. Built in the 1990s on a site adjacent to Modernist apartment blocks dating from East Germany's socialist era, Kirchsteigfeld has been developed from a master plan established by the architectural firm of Rob Krier and Christoph Kohl (Krier and Kohl 1999). Their plan ensured architectural variety by assigning different architects to the design of adjacent buildings. Most of the 2,300 housing units are social housing, supported by public subsidies. They are framed in medium-rise, high-density structures that are organized around courtyards with communal gardens, echoing in a larger and more spacious format the nineteenth-century tenement buildings of the region. The street network is punctuated by a pond and a linear water feature, and its edges are carefully landscaped and furnished with scalloped benches (Figure 5.5).

Figure 5.5 Kirchsteigfeld, Germany. The town echoes earlier high-density urban patterns, with a well-defined network of relatively narrow streets with lively facades and generous courtyards. This photograph shows the landscaped central axis, with the adjacent rondelle and Horseshoe Park surrounded by mid-rise courtyard buildings. (Photo: Photographie Werner Huthmacher, Berlin)

Figure 5.6 Horseshoe Square, Kirchsteigfeld. Each section of Kirchsteigfeld has its own public open space, each designed in a different shape and size, intended to foster the development of a very small-scale sense of place and identity. (Photo: Johann Jessen)

The generous landscaping of the town recalls garden city projects of the early twentieth century, such as Germany's Margarethenhöhe, and in fact the town's most distinctive features are its open spaces. Each section of the town has its own uniquely configured open space. The most striking of these is the teardrop-shaped park (bafflingly named Horseshoe Square) in the north part of the town (Figure 5.6). Surrounded by apartment buildings, surfaced with gravel and planted with formal rows of trees, it is reminiscent of the Place Dauphine in Paris. Paired with a small adjacent space at the centre of a circular arrangement of six-storey apartment buildings, the two spaces form an exclamation point in plan view. At street level, the tightly enclosed arena of the circular space gives way through a narrow opening to the more expansive vista of the park. More important functionally is the central market square, which has a landmark church and a grouping of retailing and commercial services and public institutions, including a community centre, a branch library, a high school an elementary school and two day nurseries.

Barely fifteen years old, to some observers the town

> still feels like a set piece, a stage set in which it is not yet obvious that the
> quality of community will match the thoughtful design of most of its constituent

pieces. Some of the beautifully landscaped communal areas seem to have been designed more for display than use.

(James-Chakraborty 2001: 60)

Preliminary responses from residents, however, indicate otherwise. There is wide-spread recognition of the importance of the public squares and the way they frame urban spaces and lend identity and a sense of place. Residents 'all cherish the attractive views the landscaped interiors offer from inside their apartments, and second, everybody recognises the possibilities for communal gatherings which these places offer' (Basten 2004: 96). The semi-public interior spaces of the blocks are used by children to move between houses, and the play areas in these spaces – safe and easy to supervise even from within the apartments – are highly valued by parents. Meanwhile, the central market square is universally identified by residents as a focal point for meeting neighbours and as important for the community as a whole, since it is the location of nearly all special events. Residents perceive both its location (especially in relation to nearby shops and public services) and its size as important.

Key reading

P.L. Knox and H. Mayer (2009) *Small Town Sustainability*. Basel: Birkhäuser, p. 92.

of national and global culture. Others have characterized neotraditionalism as a marketing strategy that plays on loyalty to an idealized nuclear family and attempts to resituate women in the role of the homemaker at a time when traditional gender roles are weakening (Leslie 1993; Veninga 2004).

Meanwhile, adherents of neotraditional design sought to seize the initiative by rebranding their ideas and principles under the banner of the 'New Urbanism'. The canon was established by Andres Duany and Elizabeth Plater-Zyberk, whose design for the Florida resort town of Seaside has become an icon of the movement. Their firm, DPZ, drew up a 'Lexicon of New Urbanism' and shared it with a newly formed (in 1993) Congress for the New Urbanism (CNU). The central tenet of the New Urbanism is that both civic architecture and pedestrian-oriented streets can act as catalysts of sociability and community. Tree-lined streets are designed with a comparatively narrow width, and lined with stoops or front porches as social buffer zones between the public realm of the street and the private realm of the home. As in TND, culs-de-sac are avoided; small lots, mixed uses and side alleys are encouraged. Towns are conceived as being made up of a series of clearly identifiable neighbourhoods and districts, with pedestrian-oriented commercial enterprises and civic spaces like schools, parks and community centres distributed throughout the neighbourhoods, and vehicular traffic routed through boulevards that provide axes of orientation. Larger commercial activities are concentrated in a town centre, along with significant civic structures such as churches and local government buildings. All this is to be achieved, according to the CNU, through detailed prescriptive codes and conventions, embedded in a series of regulatory documents – a Regulating Plan, Urban Regulations, Architectural Regulations, Street Types and Landscape Regulations. In addition, a 'SmartCode', developed and copyrighted by DPZ, allows them to be zoned incrementally along the lines of a Geddes-like urban–rural transect.

Kentlands, in Gaithersburg, Maryland, is generally regarded as the most successful of the commonly cited new urbanist developments. Nevertheless, critics have pointed out that although it does have a mix of housing types, Kentlands is basically an enclave of affluent upper-middle-class households (Kim and Kaplan 2004; Grant 2006). In Britain, the showcase example of new urbanism is the ersatz village of Poundbury, in Dorset (Figure 5.7).

The principal strength of the New Urbanism is embodied in the progressive principles embraced in the CNU's Charter: nine principles of regional-scale development; nine at the scale of the neighbourhood, the district and the corridor; and nine at the scale of the block, the street and the building. All are eminently progressive and desirable elements of good and sustainable placemaking (Ellis 2002). The principal underlying weakness of the New Urbanism, though, is one that is shared with all of its antecedents: the conceit of environmental determinism and the privileging of spatial form over social process. As Edward Robbins (2004) notes:

> The New Urbanism, like Modernism, can be accused of a kind of essentialism, in which all aspects of the complex and diverse urban world are reduced to a set of singular and authoritative principles summarized in a set of simple statements and strategic visual and verbal discourses.
>
> (Robbins 2004: 228)

In the prescriptive reasoning of the New Urbanism, this essentialism is magnified and laid bare. Design codes become behaviour codes; cultural myopia masquerades as universal values. David Harvey (2000), borrowing from Louis Marin's (1990) categorization of Disneyland, has described the packaged landscapes of master-planned communities and the New Urbanism as paradigmatic 'degenerate utopias'. Like Disneyland, they are designed as harmonious and non-conflictual spaces, set

Figure 5.7 Poundbury, Dorset, England. Built on land owned by the Duchy of Cornwall on the outskirts of Dorchester, Poundbury has been built on new urbanist principles, with neotraditional building styles. Buildings are made of local and recycled materials and built using traditional methods. (Photo: Tim Graham/Getty Images)

aside from the real world. Like Disneyland, they incorporate spectacle and maintain security and exclusion through surveillance, walls and gates; and, like Disneyland's Main Street, they deploy a sanitized and mythologized past in invoking identity and community. All of this is 'degenerate', in Harvey's view, because the oppositional force implicit in the progressive and utopian ideals embraced by the design professions has mutated, in the course of materialization, into a perpetuation of the fetish of commodity culture.

What New Urbanism has evidently got right, along with other kinds of master-planned developments, is its market appeal. But New Urbanism in practice is often a far cry from the New Urbanism of the CNU Charter. Evangelical New Urbanist consultants have been wildly successful on behalf of developers (who quickly recognized its attractiveness in branding their products). New Urbanism has become extraordinarily influential, its rosy rhetoric not only popular with developers but also with civic leaders desperate to promote upscale development. As premium spaces designed to accommodate the 'secession of the successful' (Reich 1991), New Urbanist developments are perfectly suited to the shift in social, cultural and political sensibilities that has occurred with the rise of neoliberalism. Form, after all, follows finance. In a classic case of co-optation, New Urbanism has been transmuted from a critical and potentially progressive force into an instrument of the prevailing order. Now, developers everywhere are using the tag 'New Urbanist' as a kind of designer branding for privatized dioramas and picturesque enclaves of what basically amounts to upscale sprawl.

Urban design and economic competitiveness

The increasing entrepreneurialism of urban governance has made rebuilding, repackaging and rebranding the urban landscape a common priority among large industrial cities. Flagship cultural sites, conference centres, big mixed-use developments, warehouse conversions, waterfront redevelopments, heritage sites, and major sports and entertainment complexes have appeared in many cities. Geared toward consumption rather than production, these settings are designed to provide a new economic infrastructure suited to the needs of a postindustrial economy: business services, entertainment and leisure facilities, and tourist attractions. They are, invariably, closely woven into the narratives of city branding. They are also, invariably, the product of 'growth-machine' coalitions of local real estate, finance and construction interests that seek to propagate an ideology of growth and consumption as well as engaging in tactical politics around local government land-use regulation, policy and decision-making and in pursuing public–private partnerships. According to political scientists John Logan and Harvey Molotch (1987), the principal agenda of local elites – the *rentier* class of landowners, developers, realtors or estate agents, bankers, construction companies and the auxiliary players in utility companies, engineering and technical subcontractors, retailers, chambers of commerce, local media and so on – is to secure the preconditions for growth. This involves advancing the ideology of growth and consumption: the neoliberal agenda (Swyngedouw *et al.* 2002). Political leaders are drawn in to growth-machine coalitions because local governments rely heavily on property taxes to fund infrastructure and essential services such as schools, police and fire protection. Support – often tacit support – also comes from professionals (including architects, engineers, landscape architects, planners and surveyors) whose jobs depend on growth.

An early example in the context of the neoliberal era comes from New York City, where a coalition of elites first came together to forge a new image for the city in response to a deep fiscal crisis in the 1970s. In addition to the fiscal fragility of the city government, New York was characterized at that time by civil unrest, blackouts, strikes, neighbourhood abandonment, graffiti-covered subways and soaring crime. Building on initiatives by entrepreneurs in media, real estate and tourism – notably the 'Big Apple' branding exercise by the Association for a Better New York and the efforts of the newly founded *New York* magazine to portray the city as a cool place for young urbanites to live, work and shop – Mayor John V. Lindsay organized one of the first public–private partnerships, aimed simply at sprucing up the city in time for the Bicentennial celebrations of 1976: cleaning taxicabs, sweeping the streets and handing out golden apple lapel pins at airports and train stations. In 1977 the New York State Department of Commerce started the 'I ♥ New York' campaign, made world famous by the simple graphic designed by Milton Glaser. Miriam Greenberg, in *Branding New York* (2008), argues that this was the beginning not only of an image makeover, but also of a systematic strategy of restructuring political and economic relations in the city toward a more business- and tourist-friendly environment. Reference has already been made (p. 134) to the example of the Grand Central Partnership; while Neil Smith (1996) has written about the strategic gentrification of parts of Manhattan in the cause of a 'good business climate'.

In the 1990s, Mayor Rudy Giuliani pursued an aggressive law enforcement and deterrent strategy, with highly publicized crackdowns on relatively minor offences such as graffiti and turnstile jumping on the theory that this would send a message that order would be maintained, thereby enhancing the quality of life for the city's middle classes. His successor, Michael Bloomberg, has continued the rebuilding, repackaging and rebranding of the urban landscape, supporting the construction of new sports stadia, pushing to attract and support huge events like the Republican National Convention and New York Fashion Week, establishing the office of a chief marketing officer for the city, and marketing the city under a new slogan, 'The World's Second Home'.

Urban designscapes and hard branding

Along with rebuilding, repackaging and branding have come events, exhibitions, and a renaissance of urban design in both public and private sectors. Urban design curricula have been revived or expanded in many universities as a result of the increased demand for urban design expertise by architectural and engineering firms undertaking large-scale projects. Governments, for their part, have come to focus more on the competitive advantages of managed urban design. In England, for example, the Commission for Architecture and the Built Environment was set up in 1999 to promote good architecture and urban design, while in London the Mayor's office has established an advisory group, Design for London, which answers directly to the Mayor and works closely with staff at the London Development Agency, Transport for London, the Greater London Authority and the individual boroughs within London. Design for London has commissioned an analysis and assessment of the city's public realm in order to establish an over-arching strategy and set out design criteria. Among Design for London's design projects are area master plans, infrastructure improvements, housing developments and a programme to create or upgrade public spaces.

The net result is the appearance of what Guy Julier (2005) calls 'designscapes': distinctive ensembles of new buildings, cultural amenities, heritage conservation projects, renovated spaces, landscaping and street furniture with, inevitably, an associated programme of planned events and exhibitions. Designscapes also extend, notes Julier,

> to the productive processes of design policy-making and implementation, design promotion and organisation and the systems of provision of design goods and environments within these contexts. The actor networks that make up the design culture of a location therefore take on a symbolic role – they become a kind of 'meta-activity' that frames and explains urban social, environmental and economic identity.

(Julier 2005: 874)

The designscapes of large-scale urban development projects are increasingly the source of place identity, adding a new dimension to cities' industrial specializa-tions, the traditional markers of place identity (see Case Study 5.2: Millennial regeneration in Portsmouth). One of the earliest examples in the United Kingdom was Salford Quays on the Manchester Ship Canal, initially developed in 1982

Case Study 5.2 **Millennial regeneration in Portsmouth**

With the steady reduction in size of the Royal Navy after the Second World War, the city of Portsmouth, Britain's principal naval port, was an early victim of deindustrialization. New weapons and ships meant that the size of navies could be dramatically reduced without reducing their capabilities. Whereas the Royal Navy employed more than 21 per cent of total working population of Portsmouth in the 1930s, the figure had fallen to 9 per cent by 1981. With the civilian dock-yard labour force dramatically reduced in size, and disinvestment by the Royal Navy in much of the dockyard infrastructure and its defence-related facilities, the city had taken on a rather grim aspect by the mid-1980s, with a significant amount of derelict land and an economy that was in poor shape. Unemployment was high, averaging 10 per cent, with pockets as high as 86 per cent on the Landport Estate, a part of the Charles Dickens electoral ward that consistently appears in the bottom 10 per cent on the UK Government's Deprivation Index.

As the city looked for possible strategies for regeneration, it became clear that its principal assets were its waterfront (much of which was Ministry of Defence property and so had been out of bounds to the public for many years) and the historic naval buildings and attractions, which include the restored remains of Henry VIII's flagship, the *Mary Rose*, HMS *Victory* (Nelson's flagship from the Battle of Trafalgar 1805), HMS *Warrior* (Britain's first iron-clad steamship 1860), the Royal Naval Museum, a Tudor castle, a series of ramparts and fortifications, and a D-Day museum.

Beginning in the early 1980s, a neoliberal 'growth-first' strategy was adopted by the city. A pro-growth coalition involving local authority planning and marketing departments, a prominent local architect, members of the local university and local political activists eventually led to the formation of a public–private partnership (Portsmouth and South East Hampshire Partnership: www.the-partnership.co.uk) in 1993. After a series of failed attempts to secure investment, the partnership was successful in securing £40 million of Millennium Commission funding (derived from the National Lottery) in 1995, with a strategic focus on the regeneration of Portsmouth Harbour. The scheme was based on the regeneration of the Victoria and Alfred Docks in Cape Town, South Africa, where a festival marketplace had emerged on disused dockland sites.

The centrepiece of Portsmouth's regeneration scheme is the iconic Spinnaker Tower (Figure 5.8), a 170-metre (558-ft) observation tower whose sail-like form was designed to echo the maritime branding of the city. From its three glazed viewing decks the tower offers visitors uninterrupted panoramic views across the harbour, the city and, across the Solent, the Isle of Wight. Adjacent to the tower on land that was formerly a naval base (HMS *Vernon*) is a mixed-use development, Gunwharf Quays, an extensive waterfront redevelopment scheme that contains a series of condominium buildings (dominated by the 29-storey No. 1 building, whose form resembles a ship's funnel), together with almost 100 designer shopping outlets, 25 bars and restaurants, a 14-screen cinema, a bowling alley, nightclubs, a contemporary art gallery and a 120-bed hotel. The development overlooks the Gunwharf Quays Marina, which hosts spectacular sailing events such as the Volvo Ocean Race and accommodates luxurious super-yachts and visiting tall ships.

Across the harbour to the west the Partnership's regeneration efforts have focused on two residential developments: Priddy's Hard (the Royal Navy's former armaments depot) and Royal Clarence Yard (former naval victualling yards). A feature of the Priddy's Hard development is 'Explosion!', a museum of naval warfare housed in a group of listed buildings centred around the original powder magazine

Figure 5.8 Spinnaker Tower and Gunwharf Quays, Portsmouth. The tower is visible for miles around the city, and has become a key part of the city's identity. (Photo: Jean Brooks/Robert Harding World Imagery/Corbis)

of 1777. On the eastern side of the harbour, meanwhile, historic fortifications have been cleaned up, opened to the public, and linked by a pedestrian Millennium Walk.

Altogether, these developments have created over 2,500 jobs and attracted over £300 million of investment to the area, earning the Project of the Year and Regeneration Award 2006 from the Royal Institution of Chartered Surveyors. Together with the naval museums and heritage ships and an expanded series of car and passenger ferry services from Portsmouth Harbour to France and Spain, they have given critical mass to Portsmouth's attractions as a major UK tourist destination. The shops and restaurants at Gunwharf Quays now attract over 115,000 visitors per week, and the development as a whole has been the catalyst for an accelerated gentrification of the neighbouring Portsea district.

The commercial success of the harbour regeneration has prompted other major regeneration schemes. Continuing the Partnership's vision of 'architectural renaissance', the Portsmouth Gateway Project (a range of environmental improvements and landmark features along the M275 entrance to the city) was established in 1999; and in 2004 a £500 million regeneration scheme was announced for the city's Northern Quarter, adjacent to the city's principal commercial centre. The scheme involves ten new city blocks with a hierarchy of new open streets and public spaces accommodating a mixed-use development that includes affordable housing. Originally scheduled for completion in 2011, the scheme was put on hold as a result of the global financial crisis of 2008–2009.

Key readings

Ian Cook (2004) 'Waterfront Regeneration, Gentrification and the Entrepreneurial State: The Redevelopment of Gunwharf Quays, Portsmouth', Spatial Policy Analysis, School of Geography, University of Manchester, Working Paper, 51.

Nancy Holman (2007) 'Following the Signs: Applying urban regime analysis to a UK case study', Journal of Urban Affairs, 29.5: 435–453.

through public–private partnerships on the site of Salford Docks following the closure of the dockyards. The development now includes apartment blocks, offices, hotels and retail space, together with the Imperial War Museum North (designed by Daniel Libeskind) and a landmark arts venue, the Lowry arts centre (Figure 5.9), designed by James Stirling and Michael Wilford. Other European examples include the new financial district in the Dublin docklands, the science-university complex at Adlershof in Berlin, Copenhagen's Orestaden project, and the 1998 World Expo site in Lisbon.

The largest and most important designscape of all is the redeveloped Docklands area of London (Figure 5.10). Once the commercial heart of Britain's empire, employing over 30,000 dockyard labourers, London's extensive Docklands fell into a sharp decline in the late 1960s because of competition from specialized ports using new container technologies. In 1981 the London Docklands Development Corporation was created by the central government and given extensive powers to redevelop the derelict dock areas. The remaking of the Docklands in the 1980s was a deliberate attempt by Prime Minister Margaret Thatcher's government not simply to market this part of London to global investors, but to sell the whole idea of the United Kingdom as a rejuvenated, postindustrial economy. The massive redevelopment programme has seen a huge area of the Docklands converted into a mixture of residential, commercial, exhibition and light industrial space, the largest single urban redevelopment scheme in the world (Carmona 2009b). It should be noted, however, that the Docklands scheme

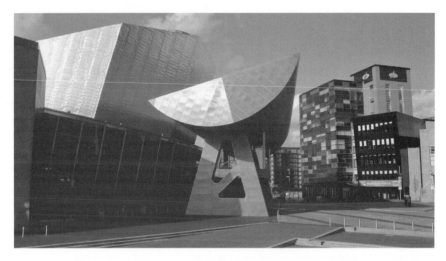

Figure 5.9 The Lowry arts centre, Salford Quays, Greater Manchester. Named for L.S. Lowry, the artist who painted urban-industrial scenes populated by 'matchstick' figures, the Lowry arts centre includes a gallery housing 350 of Lowry's paintings as well as a drama studio and two theatres. (Photo: author)

Figure 5.10 London Docklands. A massive development programme managed by the London Docklands Development Corporation during the 1980s and 1990s saw a huge area of derelict docklands converted into a mixture of residential, commercial and light industrial space, dominated by the Canary Wharf project that has become a second major financial centre in London. (Photo: Jason Hawkes/Getty Images)

prompted intense ideological debate about the directions which regeneration should take. The London Docklands Development Corporation, was widely criticized for being unaccountable, for promoting undemocratic practices, failing to deliver benefits to the poor, increasing social polarization, and encouraging gentrification (Bird 1993; Florio and Brownill 2000; Butler 2007).

Graeme Evans (2003) uses the term, 'hard-branding' to describe the impact of flagship buildings (like new stadia, museums, arts complexes, theatres or opera houses) and major events (like the Olympics and the World Cup) on urban regeneration and place identification. Without careful coordination, as Evans (2003: 417) observes, this can result in 'a form of Karaoke architecture where it is not important how well you can sing, but that you do it with verve and gusto'. As Julier (2005) notes, hard branding has provoked 'me too' strategies, resulting in the serial reproduction of would-be signature buildings by 'name' architects, ultimately homogenizing the identities of cities and attenuating any advantage accruing to their competitiveness in attracting and retaining businesses and residents.

In addition to investment in flagship buildings, some cities have sought to promote design itself as a motor for economic regeneration (Punter 2009; Tallon 2009). Examples from Britain include Dundee by Design, the Liverpool and North Manchester Design Initiative, Design Yorkshire, the Leeds Architecture and Design Initiative, and the North Staffordshire Design Initiative (Bell and Jayne 2003). The focus here is on one or more of three dimensions of design. First, high-quality detailing of public spaces: fountains, sculptures, seating, lighting, street furniture and landscaping in support of both the city's overall image and of an ordered and prosperous atmosphere that might enhance its business climate. Second, promoting and enhancing the image of a city's commitment to design, highlighting the economic impact of design activities in a postindustrial economy, fostering design talent, and providing business advice and support to design firms. Third, extending the range and capacity of design services and matching the work of creative and design firms with local manufacturing, engineering, advanced technology and other industries to encourage design-led production.

One way or another then, design has become a key attribute of economic competitiveness for many cities. In the process, it has re-emphasized the Janus-faced condition of the urban design professions and the historical contradiction between, on the one hand, planning for environmental quality and social need and, on the other, planning for competitive accumulation. In Britain, New Labour's Urban Task Force, set up under the leadership of architect Richard Rogers in 1997, declared that it would 'identify causes of urban decline' and 'establish a vision for urban regeneration founded on the principles of design excellence, social wellbeing and environmental responsibility within a viable economic and legislative framework' (Urban Task Force 1999: 2). But, in practice, in the dominant neoliberal climate, planning and design have been increasingly subordinate to private developers. Tellingly, the Task Force itself asserted that 'One of the most efficient uses for public money in urban regeneration is to pave the way for investment of much larger sums by the private sector' (Urban Task Force 1999: 23). In this, at least, it has succeeded: neoliberal-style urban regeneration has created luxury housing, cultural amenities, new consumption spaces for the affluent middle classes and (an oversupply of) modern office space. But this has resulted in few new jobs, while the renewed emphasis on signature structures and aestheticized urban design masks the underlying problems of poverty and inequality, of affordable housing, and of inadequate and deteriorating infrastructure and educational and health care facilities. Citing the example of Glasgow, Alex Law and Gerry Mooney (2005) argue that:

> Glasgow's widely acclaimed culture-led regeneration programme is the pre-eminent example of enticing back middle class consumers and service-sector jobs to restore the fortunes of a city beset by long-run industrial decline, unemployment and slum housing. Glasgow shed 197,000 manufacturing jobs and acquired 145,000 in services between 1971 and 2001, and now has a much lower proportion of manual workers than the UK as a whole. It has become the subject of an incessant marketing campaign, emphasising art, culture and architecture, designer shopping and luxury apartments in the restored bourgeois residential quarter, the Merchant City. Last year [2004] saw the risible promotion of the city as Scotland's fashion answer to Milan: 'Glasgow: The New Black' or 'Glasgow: Scotland with Style'. Left behind are the working class of Glasgow's large peripheral housing estates, which are in an acute state of decay.
>
> (Law and Mooney 2005)

The conclusion is that the 'Glasgow model' of culture-led regeneration has contributed to the worsening levels of poverty and deprivation and to the deepening inequalities that characterize the city today:

> It has done this primarily by constructing Glasgow's future as a low paid workforce, grateful for the breadcrumbs from the tables of the entrepreneurs and investors upon which so much effort is spent attracting and cosseting – and by marginalising and ruling out any alternative strategy based upon large-scale public sector investment in sustainable and socially necessary facilities and services.
>
> (Mooney 2004: 337)

Design for the Dream Economy

But for those with money, the neoliberal era saw conspicuous consumption reach new levels. The driving force was the baby-boom generation, which had engaged briefly with radical progressivism at its coming-of-age in the late 1960s, only to form the bulwark of voter support for neoliberalism after the unpleasant shock of a sharp economic recession in the mid-1970s, when baby-boomers found themselves flooding labour and housing markets just as the economy was experiencing the worst recession since the 1930s. Salaries stood still while house prices ballooned. In 1973,

> the last really good year for the middle class, the average 30-year-old man could meet the mortgage payments on a median-priced house with about a fifth of his income. By 1986, the same home took twice as much of his income. In the same years, the real median income of all families headed by someone under 30 fell by 26 percent.
>
> (Butler 1989: 77)

It prompted apostasy:

> In our 20s, my friends and I hardly cared. We finished college (paid for primarily by our parents), ate tofu, and hung Indian bedspreads in rented apartments. We were young; it was a lark. We scorned consumerism. But in our 30s, as we married or got sick of having apartments sold out from under us, we wanted nice things, we wanted houses.
>
> (Butler 1989: 77)

An entire generation shifted its focus from radical idealism to self-oriented materialism.

In the 1980s, the lifestyles of the rich and famous, portrayed in detail in popular magazines and television programmes, became a source of inspiration as well as titillation to the merely affluent and vacuous. The upper-middle classes began conspicuously acquiring luxury goods and symbols in the 1980s – McMansions, designer clothes, shoes and bags, luxury vehicles, and prestige watches and pens. Bolstered by trends in financial markets and the escalating salaries of the new economy, the affluent middle classes let rip with competitive luxury spending. The consumption binge – and the 'self-illusory hedonism' of the 'Dream Economy' was sustained by a credit industry that became increasingly indulgent about extending credit and increasingly generous in how much it would let consumers borrow, as long as those customers were willing to pay high fees and risk living in debt.

Credit gave so many consumers access to such a wide array of high-end goods that traditional markers of status lost much of their meaning. In their place came an increasingly crude calculus of status: the bigger, the newer, the more expensive, the flashier, the better. Lifestyle magazines like *Architectural Digest*, *Elle Décor*, *Flaunt*, *GQ*, *Living*, *Luxe*, *More*, *Self* and *Wallpaper**, along with 'reality' television, amplified this conspicuous, competitive consumption. In the United States, the appetite of impulse buyers watching the Home Shopping Network was so great that the network's call centres grew to some 23,000 incoming phone lines capable of handling up to 20,000 calls a minute (Twitchell 2002). Even after the dot.com stock-market bubble of the late 1990s burst, the Dream Economy continued full flight, the credit habit deeply engrained with consumers. Inevitably, house prices soared; but mortgage financiers found increasingly novel ways to package loans to help people afford houses that otherwise would be beyond their reach, overheating the market and eventually contributing to the global financial meltdown of 2008–2009. It was a case of 'affluenza' (De Graaf *et al.* 2001).

In this supercharged economic and cultural climate, design came to assume an ever-more important role. Commercial success, as Patrick Jordan has noted, has become increasingly dependent on being able to design 'pleasurable products' (Jordan 2002). It has also become increasingly dependent on pleasurable spaces and places: the physical settings for shopping, recreation, living and working. The cachet of 'design' has come to be applied to product lines of every description, often through the 'brand extension' of famous architects and (especially) couturiers. Fashion houses have expanded strategic brand management from couture to mass-market clothing as well as cosmetics, perfumes, jewellery and accessories. Some fashion brands have diversified into product areas that have no real connection to clothing or personal attire: Armani hotels, cafés, chocolates, mobile phones, nightclubs and florists; Bulgari hotels, restaurants and resorts; Dolce & Gabbana spas, restaurants and mobile telephones; Gucci cafés; Prada mobile phones; and Versace watches, soft furnishings, crockery and hotels (Leslie and Reimer 2003a; Power and Hauge 2008). Amid this heightened sensitivity to design and the hedonism of the Dream Economy, design itself became unmoored from the dominant Modernist ideology, with a variety of impulses emerging in what came to be called 'postmodern' design.

Postmodern design

The term 'postmodern' implies a clear and decisive shift from Modernism and modernity, but in terms of design it can best be thought of as describing anything and everything that Modernism was not. In architecture, the beginnings of postmodern design were explicit, stemming from the work of architect Robert Venturi (1966), who openly challenged Ludwig Mies van der Rohe's famous Modernist aphorism, 'Less is More'. 'Less is a bore', argued Venturi; hybrid elements should displace Modernist purity, witty and ironic references should be included in design, and complexity and contradiction should have a place. In short, architects should 'Learn from Las Vegas'. Venturi was really arguing for an architecture appropriate to a new phase of urbanization, characterized by increasingly materialistic societies and widespread misgivings about the ability of modernization to deliver progressive economic and social outcomes. Whereas the architecture and urban planning of the Modernist city reflected a striving for progress, postmodern architecture and urban design plays to consumption, hedonism and the generation of profit with little regard for the social consequences. Among the most influential examples of postmodern architecture are Michael Graves' Portland Public Service Building in Portland, Oregon; Philip Johnson's Sony Building (originally the AT&T Building) in New York City; Robert Venturi's Sainsbury Wing of the National Gallery in London; the Stata Center on the MIT campus by Frank Gehry (Figure 5.11); Charles Moore's Piazza d'Italia in New Orleans (Figure 5.12); the Humana Building in Louisville by Michael Graves; and the Haas Haus in Vienna by Hans Hollein.

Figure 5.11 **The Stata Center, Cambridge, Massachusetts. Designed by Frank Gehry for the Massachusetts Institute of Technology, the Stata Center houses MIT's Computer Science and Artificial Intelligence Laboratory, the Laboratory for Information and Decision Systems, and the Department of Linguistics and Philosophy. (Photo: Angus Oborn/DK Limited/Corbis)**

Figure 5.12 Piazza d'Italia, New Orleans. Designed in 1978 by Charles Moore for the city's Italian community and intended to symbolize the revitalization of Italian culture, the Piazza was one of the first exercises in combining classic and modern motifs. (Photo: author)

Since the mid-1970s, postmodernism has permeated every sphere of creative activity, including art, advertising, philosophy, clothing design, product design, interior design, graphic design, music, cinema, novels and television, as well as architecture and urban design. Examples of postmodern product design include the work of Aldo Rossi, Frank Gehry, Michael Graves and Morphosis for Alessi; while postmodern furniture design is exemplified by the unconventional materials, historic forms, kitsch motifs and gaudy colours of Ettore Sottsass and the Milan-based Memphis group (Collins and Papadakis 1990). In postmodern graphic design, anything goes, especially overt challenges to the conventions that were once widely regarded as constituting good practice (Poynor 2003). In contrast to the purity, unity and order of Modernism, postmodern design seeks to express the exact opposite: messy vitality, hybridity, ambiguity and inconsistency. In contrast to the abstract formalism of Modernism, postmodern design aims to be decorative and scenographic, full of signs and symbols, wide-ranging and eclectic. One of its principal tropes is 'double coding', combining one kind of style or motif with another. This mixing allows the deployment of the symbolism of everything from historicism and revivalism to metaphysical references and kitschy pastiche. Postmodernism is a style of styles. This self-conscious stylishness lends itself easily to commodification; but it has not displaced Modernism, which continues to appeal to particular class fractions and market segments (especially, ironically, affluent and fashion-conscious metropolitans, for whom a sort of luxo-minimalism is the preferred high aesthetic).

Developers soon realized that although it might cost more to build a 'rich' building in postmodern style, it would likely sell or rent more quickly – and often at a premium – because it would facilitate product differentiation and project an appropriate 'look' for the consumption-oriented market segments of the 1980s and 1990s. As a result, commercial and residential townscapes throughout metropolitan areas have been suffused with the hallmarks of postmodern design. New single-family homes have been built in revival styles or historicist modes, while new stores and offices have been plastered with arches, atria, columns, keystones, semicircular windows and cornices; and office villages and shopping centres have come to resemble period stage settings: 'lite' architecture, the built environment's equivalent of easy-listening music. Postmodern design has become, to borrow David Harvey's term, the 'cultural clothing' of the neoliberal era.

The double-coding and hybridity of postmodern design is seen by many as both cause and effect of an increasing importance of signs, images and reflexivity in everyday life: something that amounts to a significant cultural shift, a postmodern sensibility, in fact. This has given rise to an extensive theoretical and philosophical literature that has influenced all of the social sciences (Lyotard 1984; Harvey 1989b; Jameson and Fish 1991). For the most part, though, the designers of post-modern architecture, graphics and products have operated independently of such philosophical sophistry.

Brandscapes of the new economy

The term 'brandscapes' has been popularized by Anna Klingmann (2007), whose book begins by noting that we have arrived at a stage of hypercapitalism 'where counterculture has been demystified, culture hijacked to transport commercial messages, (and commerce hijacked to transport culture), and all boundaries between high and low design, concept, content, and form have been blurred' (Klingmann 2007: 1). Klingmann's response, paraphrased, is that if you can't beat them, join them. She argues, rather quixotically, that architecture can make con-structive use of branding strategies in order to 'release architecture from its formal, aesthetic, and moralizing corset' (Klingmann 2007: 3) and allow it to create affirmative spaces that help generate positive affect: a deliberate construction of context, rather than the production of static and discrete sculptural objects. In exploring the prospects for such an approach, however, Klingmann allows that existing brandscapes are very much a product of corporate interests, the conjunc-tion of economic globalization and the increasing exteriorization of corporate identities:

> Brandscapes constitute the physical manifestations of synthetically conceived identities transposed onto synthetically conceived places, demarcating

> culturally independent sites where corporate value systems materialize into
> physical territories. . . . Today, more than ever, brandscapes as physical sites
> have become key elements in linking identity, culture, and place.
>
> (Klingmann 2007: 83)

Examples of such brandscapes include Times Square in New York City and
Potsdamer Platz in Berlin, both redeveloped through public–private partnerships
in the early 1990s. Times Square is now strongly linked to the identity of the Disney
corporation, which has built in an edited and sanitized way on the history of
the locality as a vibrant (though at times rather sketchy) entertainment district. It
is dominated by 'New York Land', a themed shopping area split among three
properties owned by the Disney, Warner Brothers and Ford corporations. The
redevelopment was master-planned by Robert A.M. Stern, who sought to devise
a scheme of 'different scales, signs, styles and historical periods . . . that will look
noisy, historically layered . . . unplanned . . . bold and brash' (quoted in Klingmann
2007: 84). A classic postmodern pastiche, in other words. Similarly, Potsdamer
Platz, redeveloped to a plan by architect Renzo Piano, became strongly linked
to the identities of Daimler and Sony (though Daimler sold its nineteen buildings
at the heart of Potsdamer Platz to the Swedish banking group SEB in the wake of
a difficult separation from its former subsidiary, Chrysler, in 2007). The strategy,
as in Times Square, was to commodify the legend of a world-famous site, attaching
it to corporate brand images.

A more common form of brandscape is the high-end shopping district, typically
colonized in larger cities by the flagship stores of the leading global brands of
high-end ready-to-wear clothing, accessories, jewellery, shoes and so on, supported
by expensive restaurants, cafés, art galleries, antique shops and specialized luxury
retail stores like Cerruti, Coach, Fendi, Ferragamo, Furla, Marc Jacobs, Missoni,
Moschino, Prada and Valentino. Ilpo Koskinen calls these districts 'semiotic neighbourhoods' because they specialize in selling semiotic goods and experiences: the
signifiers of distinction and cultural capital that have become so important to
the new class fractions of the new economy (Koskinen 2005). A good example
is the Augustinergasse district in Zurich in the streets adjacent to the traditional
high-end shopping street of Banhofstrasse (Figure 5.13). As in other semiotic
neighbourhoods, a street-level brandscape of global corporate identities has been
grafted onto the built environment: in this case, a well-preserved historic district
that lends a ready-made ambience for consumption as both performance and
experience as well as acquisition.

Another form of brandscape, increasingly ubiquitous, is to be found in major
airports, where concourses have been extended and remodelled to accommodate
the duty-free outlets of the same global brands that show up in semiotic neighbourhoods. Still another is the branded 'tourist trap' exemplified by Nike Town

Figure 5.13 **The Augustinergasse 'semiotic' district in Zurich, Switzerland. The
backstreets of Zurich's old town, near the financial district and
the city's principal shopping street, Banhofstrasse, are filled with a
mixture of upscale independent boutiques and branches of global-
brand specialized luxury retail stores. (Photo: author)**

developments (Sherry 1998) and by University City's City Walk in Los Angeles, a four-block-long mixed-use pedestrian promenade of shops, cafés and entertainment.

Lower down the retail hierarchy are the brandscapes of shopping malls, where both global- and national-brand stores are ensconced in a retail ecology that is supported by ample parking, multiscreen cinemas, outdoor plazas, food courts, public art works, night-time bars and maybe a rock-climbing wall. Describing the 240-acre Bluewater mall in Kent, England, Iain Borden (2000) writes:

> This is the Utopia of late capitalism, a place where all that is troublesome in the city is erased, where there are no homeless people, wailing sirens, or speeding couriers, but where there is always, with absolute 100 per cent certainty, a place to sit down, a drink to be quaffed, a toilet to be found, and a new product to be purchased. . . . This is contented consumerism where the visitor is always relaxed enough to open their wallet. . . . It is not, however, just the impressive list of facilities that produces the mood of calm and continuous spending. It is also the quality of the architecture, for Bluewater proudly displays an artful blend of wide concourses, marbled surfaces, historical styles, large sculptures, variegated colours and playful light. Barrel-vaulted roofs are interspersed with splendorous arched windows and centred oculi, floors proffer depictions

of the River Thames, cornices are decorated with representational friezes and poetic inscriptions, and the three corner-hubs contain thematized installations relating to the moon, tides, and other such uplifting conceptual armatures. This is what master architects Eric Kuhnes Associates call 'architectural diversity'.

(Borden 2000: 14)

In smaller towns, high streets and main streets have become brandscapes where independent stores and local pubs and cafés have been displaced by the dominance and homogenizing influence of national and international corporate chains. Big superstores and chain retailers have been allowed to spread by planners, town councils and governments desperate to sustain the local tax base. But the chains have become the economic equivalent of invasive species: voracious, indiscriminate and often antisocial. In small towns it does not take long for superstores, supermarkets and cloned shops to dominate and suffocate the local economic ecosystem. Their big, centralized logistical operations drive the homogenization of business, shopping, eating, farming, food, the landscape, the environment and people's daily lives.

Town centres once filled with a thriving mix of independent butchers, newsagents, tobacconists, bars, book shops, greengrocers and family-owned general stores are fast being filled with standardized supermarket retailers, fast-food chains, mobile phone shops and the downmarket fashion outlets of global conglomerates (Knox and Mayer 2009). The homogenization of the retail environment in small towns in Britain has been documented by the New Economics Foundation (Simms *et al.* 2005), which has identified 'clone towns' – places 'where the individuality of high street shops has been replaced by a monochrome strip of global and national chains, somewhere that could easily be mistaken for dozens of bland town centres across the country' (Simms *et al.* 2005: 2).

In contrast, a 'home town' is a place that retains its individual character and is instantly recognizable and distinctive to the people who live there as well as those who visit. Of the 103 towns surveyed by the New Economics Foundation, only 34 per cent had sufficient numbers of locally owned businesses to qualify as home towns, 41 per cent had so few independent stores that they were tagged as clone towns, and the rest, 26 per cent, were somewhere in between. Examples of clone towns include Stafford, Winchester (Figure 5.14) and Burton-on-Trent.

Not all of the impulses of the neoliberal era have been regressive. Liberal and progressive elements of society have focused increasingly on environmental issues, prompted not only by concerns over climate change but also by issues such as environmental justice, energy costs and a basic concern for environmental quality.

Figure 5.14 Winchester, England: a 'clone' town. The town centre was redeveloped in the 1950s and 1960s, marking the beginning of the invasion of branches of national retail chains. (Photo: author)

One response has been the emergence within the design professions of 'green' and sustainable approaches to products and places. It is, of course, something that is highly vulnerable to fashionability and commodification, with market researchers targeting 'green' consumers with products that may not in fact be as environmentally friendly as advertised. Green design and design for sustainability are taken up in detail in Chapter 8.

Summary

Contemporary cities, mostly a product of the political economy of the manufacturing era, have been thoroughly remade in the image of consumer society. Under the radically different policy perspectives of neoliberalism, a series of property booms has added significant new spaces to cities. The competitiveness of property development, combined with the increasing entrepreneurialism of city governments and the increasing materialism of popular culture, meant that development schemes have taken place at ever-larger scales, with mixed-use complexes and waterfront redevelopments, often built as public–private partnerships, offering integrated settings for mutually supporting, revenue- and tax-generating packages of retailing, offices, residences, hotels and entertainment functions. The nature of these developments, meanwhile, has blurred the traditional distinctions between public and private spaces.

Another characteristic feature of urban change during the neoliberal era has been the gentrification of selected inner-city neighbourhoods. Developers of new subdivisions, meanwhile, have catered to a different market niche by producing private, master-planned subdivisions laid out with carefully researched packages of amenities and themed settings that are matched to the finances and aspirations of different income and lifestyle groups. The design and marketing of these packaged landscapes has been strongly influenced by discourse in architecture and planning about limiting sprawl, fostering community, civility and sense of place in compact, mixed-use, walkable and relatively self-contained developments. This has drawn on the legacy of ideas that can be traced to the intellectuals' utopias of the nineteenth century, but it has often found expression in highly commodified and often regressive – rather than progressive – form.

Over this same period, globalization has rendered cities increasingly interdependent, intensifying competition among cities in attracting investment. The increasing entrepreneurialism of urban governance has made rebuilding, repackaging and rebranding the urban landscape a common priority. Flagship cultural sites, conference centres, big mixed-use developments, warehouse conversions, waterfront redevelopments, heritage sites and major sports and entertainment complexes have appeared in many large industrial cities. Geared toward consumption rather than production, these settings are designed to provide a new economic infrastructure suited to the needs of a postindustrial economy: business services, entertainment and leisure facilities, and tourist attractions.

In this radically different new economic and cultural climate, design came to assume an ever-more important role. Whereas the architecture and urban planning of the Modernist city reflected a striving for progress, postmodern architecture and urban design played to consumption, hedonism and the generation of profit, with little regard for the social consequences. Combined with the globalization of retailing and popular culture, the result has been the widespread proliferation of 'brandscapes' and cloned commercial settings.

Further reading

Helen Castle (ed.) (2000) *Fashion + Architecture*. Chichester: Wiley-Academy. A series of essays that take a close look at the relationship between fashion and architecture.

Alexander Cuthbert (2006) *The Form of Cities: Political Economy and Urban Design*. Oxford: Blackwell. A synthesis of recent thinking about urban design and a powerful analysis of the emergence, logic and political meaning of the built environment in different historical contexts.

Nan Ellin (2006) *Integral Urbanism*. New York: Routledge. Challenges the escapist and reactive tendencies of modern and postmodern urban design and planning, making the case for an 'integral' approach involving the tasks that planners and architects typically conceive of as being separate from each other.

Jill Grant (2006) *Planning the Good Community: New Urbanism in Theory and Practice*. New York: Routledge. Grant takes a critical look at how new urbanism lives up to its theory in its practice, and asks whether new urban approaches offer a viable path to 'community'.

Jason Hackworth (2007) *The Neoliberal City: Governance, Ideology, and Development in American Urbanism*. Ithaca, NY: Cornell University Press. Shows how neoliberal policies are having a profound effect on the nature and direction of urbanization.

Anna Klingmann (2007) *Brandscapes: Architecture in the Experience Economy*. Cambridge, MA: MIT Press. Shows how branding and experience management are at the forefront of contemporary architectural theory and practice, and suggests that commodified desire may give designers more control over business plans and urban politics.

Paul Knox (2008) *Metroburbia, USA*. New Brunswick, NJ: Rutgers University Press. Explores how extreme versions of the American Dream have changed the landscape of US cities.

PART III
Designer cities

In Part III, the focus shifts to cities themselves as settings for design and design services, looking at the role of design in the economy and material culture of contemporary cities, at design as a key component of urban experience, and at the urban ecologies of design as an economic activity. As new developments change the spatial grammar of power and respond to the demands generated by new cultural impulses, design – good, bad and indifferent – plays an important role in the feelings, emotions and moods evoked in different settings. These feelings, in turn, have become central to the 'experience economy' that is characteristic of contemporary cities. Meanwhile, the design professions themselves have not only grown in size and importance but also begun to reflect distinctive urban geographies, locating disproportionately in the cities most intimately connected with global commodity chains and networks of business services. The focus here is on the globalization of design services and the anatomy of designer quarters in global cities, where professional milieux are characterized by complex linkages among designers, producers and distributors.

6 Design and affect in urban spaces

The emphasis in this chapter is on the importance of the distinctive sense of place associated with particular localities and on people's collective emotional responses to the built environment, to each other, and to their community. A distinction is made between the social construction of urban space (the way people give meaning to places and spaces) and the social production of urban space (the way society is organized to produce and modify the built environment). Both have been influenced by globalization, by the trend toward increasing consumerism, and by people's growing appetite for spectacle and experience. Three outcomes are illustrative of the way that these trends have impacted cities: the commodification of urban nightlife, the proliferation of 'starchitecture' and the 'museumization' of central city districts. Many cities have adopted strategies of brand management and urban regeneration through the promotion of culture and design, following the success of Barcelona and Bilbao.

In societies where the commonalities among places are intensifying, the experience of distinctive places, physical settings and landscapes has become an important element of consumer culture. Vast tracts of large cities in Europe and North America, with their transnational architectural styles, dress codes, retail chains and popular culture, are associated with feelings of placelessness and dislocation, a loss of territorial identity, and an erosion of the distinctive sense of place associated with particular localities. But it is not simply a matter of the aesthetics of the built environment. Sense of place is always socially constructed, and a fundamental element in the social construction of place is the existential imperative for people to define themselves in relation to the material world. The roots of this idea are to be found in the philosophy of Martin Heidegger, who contended that men and women originate in an alienated condition and define themselves, among other ways, through their socio-spatial environment. As noted in Chapter 1, people generate meanings about objects, buildings and spaces through routinized, recursive behaviours and practices in their particular lifeworlds, the taken-for-granted context for their everyday living. Often this carries over into a collective and self-conscious 'structure of feeling', including the 'affective' dimension of feelings, emotions and moods evoked as a result of the experiences and memories that people associate with a particular place.

The basis of both individual lifeworlds and the collective structure of feeling is intersubjectivity: shared meanings that are derived from the lived experience of everyday practice. Positive affect and distinctive sense of place stem in large part from the routine encounters, shared experiences and daily rituals of urban streetlife that make for intersubjectivity. From a design perspective, this requires plenty of opportunities for informal, casual meetings and gossip; friendly bars and pubs and a variety of settings in which to purchase and/or consume food; street markets; pedestrian-friendly settings, and a variety of comfortable places to sit, wait and people-watch.

People's 'creation' of space can also provide them with roots – their homes and localities becoming biographies of that creation. Central to Heidegger's philosophy is the notion of 'dwelling' – the basic capacity to achieve a form of spiritual unity between humans and the material world. Through repeated experience and complex associations, our capacity for dwelling allows us to construct places, to give them meanings that are deepened and qualified over time with multiple nuances. Heidegger anticipated the effects of rationalism, mass production, standardization and mass values on people's capacity for 'dwelling' and the social construction of place. The inevitable result, he suggested, is that the 'authenticity' of place is subverted. Streets and neighbourhoods become inauthentic and placeless, a process that is, ironically, reinforced as people seek authenticity through professionally designed and commercially constructed spaces and places whose invented traditions, sanitized and simplified symbolism, and commercialized heritage all make for convergence rather than spatial identity.

Globalization has meanwhile prompted communities in many parts of the world to become much more conscious of the ways in which they are perceived by tourists, businesses, media firms and consumers. As a result, places are increasingly being reinterpreted, reimagined, designed, packaged and marketed. Sense of place can become a valuable commodity through place marketing. Seeking to be competitive within the global economy, many places have sponsored extensive makeovers of themselves, including the creation of pedestrian plazas, cosmopolitan cultural facilities, festivals, and sports and media events – the 'carnival masks' and 'degenerative utopias' of global capitalism (Harvey 2000).

Placemaking, in this context, becomes an inherently elite practice, determined by those in control of resources. Setha Low makes the useful distinction between the social construction of urban space and the social production of urban space. She defines *social production* 'as the processes responsible for the material creation of space as they combine social, economic, ideological, and technical factors', while the *social construction* of space 'defines the experience of space through which "people's social exchanges, memories, images, and daily use of the material setting" transform it and give it meaning' (Low and Lawrence-Zúñiga 2003: 21). Another useful observation here comes from the writings of Michel de Certeau (1984), who has shown how the meanings inscribed or imposed through the social production of space – i.e. through architecture, planning and urban design – can be subverted and transformed by people's 'spatial practices' and 'tactics'.

The affective dimensions of space

City spaces, then, are shot through with complex layers of intersubjective meanings and affect – people's collective emotional responses to their environment, to each other, and to the patterns of local economic, social and cultural activities. Affect is often overlooked because it is, by definition, taken-for-granted and difficult to quantify or classify. Affect is generated not only from the meanings ascribed to buildings and spaces, but also from the reassurance of the daily rhythms of street life – the opening of shops, the sweeping of pavements, the unloading of delivery trucks, and the banter of groups of children on their way to school; from the conviviality of a farmers' market; from the happy shouts of children at play; from the way that changing light picks out unexpected features and casts improbable shadows; and from the patterns of dress and comportment associated with changing seasons. Affect can be negative, too: generated by the rumble of traffic and the sound of police sirens, the antisocial behaviour of groups of teenagers, or the depressing air of a derelict factory.

It follows that affect is fleeting, contingent, and very difficult to pin down in either an empirical or theoretical sense. The topic has received increased attention in human geography and other social sciences in recent years (Pallasmaa 1996; Massey 2005; Anderson 2006), but as Peter Kraftl and Peter Adey (2008: 215) observe, 'affect is often discussed in relatively vague and ephemeral ways that revolve around the subject of emotion in one way or another'. Rather, they suggest, 'affect can be understood as the property of relations, of interactions, of events: It is not purely the property of a single (human) being'. Nigel Thrift (2004: 64) suggests that affect is 'a sense of push in the world', something that impels people to feel, think or act in a certain way. Thrift is especially interested in the notion of 'performance' and its relation to the affect of urban spaces (Thrift and Dewsbury 2000).

There are several dimensions to this: the reiteration of social norms as reflected, for example, in the 'performed' nature of gender identities in public spaces; the general flow of people's routines (performances) in everyday life; as well as the performative impulse that is associated with certain kinds of urban spaces, especially entertainment zones. Joseph Roach describes the latter as places where the

> magnetic forces of commerce and pleasure suck the willing and unwilling alike. Although such a zone or district seems to offer a place for transgression, for things that couldn't happen otherwise or elsewhere, in fact what it provides is far more official: a place in which everyday practices and attractions may be legitimised, 'brought out into the open', reinforced, celebrated or intensified.
>
> (Roach 1996: 27)

The importance of affect is also something that is increasingly recognized in professional contexts. The semiotics of goods and services depend heavily on advertising and marketing that deploys exclusivity, narcisissm, and totemism to evoke specific affective experiences and mood settings (Schmitt 1999). With the eclipse of pure Modernism, architects and industrial designers – especially practitioners of postmodern design – have paid increasing attention to the hedonistic as well as the functional qualities of products, buildings, and settings. As Guy Julier (2000) points out, one of the tasks of all designers is the construction of an 'aesthetic illusion' around the product. 'This means that they mediate the gap between production, in its crude form as the system for the origination and creation of goods and services, and consumption, as the user's engagement with them' (Julier 2000: 55). The danger here for designers, of course, is the recurring conceit of design determinism.

Design and the experience economy

Contemporary economies, as Amin and Thrift (2007) observe, are 'organized around the lifeworld of passions, moral sentiments, practical knowledge, modes of discipline and measurement, and symptomatic narratives' (Amin and Thrift 2007: 157). Culture and economics, as they stress, are inseparable. For some observers, affect is a key component of the cultural dimension of contemporary economies. In particular, Joseph Pine and James Gilmore (1999) write of the emergence of an 'experience economy'. Over the past 200 years, they argue, there has been a shift from an agrarian economy based on extracting commodities, to an industrial economy based on manufacturing goods, to a service economy based on delivering services and, now, to an economy where consumer experiences are increasingly the locus of profitability.

As profitability has become more challenging in agrarian, industrial and service enterprises, so businesses have come to orchestrate memorable events for their customers, with memory itself becoming the 'product' – the experience. It is not what is sold that characterizes the experience economy, but rather the way it is sold. Experience and affect thus become competitive advantages for products and services. For architecture in the experience economy, 'the relative success of a design lies in the sensation a consumer derives from it – in the enjoyment it offers and the resulting pleasure it evokes' (Klingmann 2007: 19).

The importance of experiential dimensions of the economy is by no means new. Think, for example, of sports entertainment, arena rock concerts and visits to museums and galleries, as well as film, music, dance and cuisine – all entwined in complex ways in the economy. But as more affluent consumers satisfy more and more of their needs and wants in terms of commodities, products, and services, businesses are increasingly combining these things with compelling experiences – experiences that must be designed and developed.

Restaurants organize their services around particular themes (providing marketers with a new term: 'eatertainment'). Shops and malls organize shows, events or expositions ('shoppertainment'). Education is not immune ('edutainment'). Nor is religion: in the United States, megachurches are designed to colonize every aspect of life, with round-the-clock activities that include aerobics classes, social services, fast food, bowling alleys, sports teams, aquatic centres with Christian themes and multimedia bible classes. A typical example is Radiant, a megachurch in the Phoenix exurb of Surprise. Its purpose-built 55,000-square-foot church looks more like an overgrown ski lodge than a place of worship. The foyer includes five 50-inch plasma-screen televisions, a bookstore and a café with a Starbucks-trained staff making espresso drinks. Doughnuts are served at every service, and for children there are also video game consoles. Megachurches provide a very distinctive set of experiences (and also function in somewhat parallel fashion to Islamic madrassas, sustaining and extending habitus in the fields of religion, civics and neighbourhood).

Other examples are more prosaic. There are now dental practices and general medical practices, for example, that double as day spas, while some manufacturers have established flagship stores that have a significant experiential component: Nike Town, SegaWorld and Warner Studio 'Villages', for example. In addition, some businesses are based entirely on selling new kinds of experiences. American Girl Place, for example, with branches in Atlanta, Boston, Chicago, Dallas, Los Angeles and New York, offers a combination of attractions, events and experiences specifically targeted at young girls: a doll hair salon, a special café, a theatre featuring musicals, and shops with dolls and outfits. Their website advertises 'Café. Theater. Shops. Memories'.

As Anna Klingmann (2007) and Anne Lorentzen and Carsten Hansen (2007) have pointed out, in the experience economy it is more than ever possible to capitalize on places: 'The experience economy is place bound, because of its particular characteristic of arousing feeling, forming identity and involving the customer in a more or less absorbing experience.' Most widespread, perhaps, are themed suburban communities and themed restaurants. But perhaps the best example in this context is the designerly affect of upscale shopping districts, where consumers can, in Bourdieu's terms, leverage social distinction by deploying their cultural capital while 'acting out' shopping and perhaps engaging in a bit of modern-day *flânerie* in between (Gregson *et al.* 2002).

Increasingly, it is not only the goods bought in shops that say something about who we are, or would like to be, but also the design of the shops themselves. The luxo-minimalist interior design of high-end stores – as magazines like *Wallpaper** make perfectly clear – is currently the preferred setting for the performativity of shopping. As Jane Rendell (2000) points out, the most sought-after and successful designs for high-end retail settings in this genre are often those that seem to challenge the status quo. By being a bit 'edgy' they are able to create a distinctive site and identity for a product. But the role of the radical architect or interior designer in this context requires a certain degree of complicity: 'By providing a memorable place, architecture enhances product identity, but is itself commodified' (Rendell 2000: 11). This, of course, is rarely a concern for professional designers, since this commodification, in turn, intensifies their brand image and their market value. The exemplar here is Rem Koolhaas and his store designs for Prada in New York and Beverly Hills.

Spectacle and enchantment

The experience economy can be thought of as part of the Dream Economy described in Chapter 1, intimately co-dependent with the 'romantic capitalism' and 'self-illusory hedonism' of competitive consumption. It is also a result of the need for the serial enchantment and re-enchantment of products, as noted by George Ritzer and described in Chapter 1. Ritzer (2005), following Baudrillard, Debord and others, points to the importance, in contemporary material culture, of spectacle, extravaganzas, simulation, theming and sheer size, and argues that they are all key to enchantment and re-enchantment in the consumer world. For Guy Debord (1967), spectacle was an overarching concept to describe consumer society, including the packaging, promotion and display of commodities and the production and effects of all media. The subsequent growth of digital technologies has extended and amplified spectacle as a means of promotion, reproduction, and the circulation and selling of products, while political and social life is also shaped more and more by media spectacle and tabloidized infotainment. As a result, spectacle itself has become one of the organizing principles of the economy, polity, society and everyday life (Kellner 2005).

By any account, the capital of urban spectacle and simulation must be Las Vegas, where the Strip is studded with buildings cribbed from the skylines of other cities. The landmarks of Paris are just across the street from the canals of Venice, and right down the block from the Brooklyn Bridge and the Statue of Liberty. Within the fantastical architecture of themed casino hotels – a massive black pyramid, a Disneyesque medieval castle, and so on – are spectacular circus acts, concerts and exhibitions. The Mirage, one of the world's largest casinos, has an ecologically 'authentic' tropical rain forest in a nine-storey atrium; a 20,000-gallon marine fish tank behind the reception desk; and, outside, a 54-foot volcano that spews steam and flames into the night sky for three minutes every half-hour. In the 1970s we were urged to 'learn from Las Vegas' for the symbolism of its architectural forms, for the malleability of its visage, and for its significance as 'The Great Proletarian Cultural Locomotive' (Venturi *et al.* 1977). In fact, what other cities learned from Las Vegas was the power of the carnivalesque in economic redevelopment and the importance of novelty and luxury.

More generally, the shift to a society of the spectacle involves a commodification of previously non-colonized sectors of everyday social life. In this vein, Douglas Kellner (2005) points out that in addition to the obviously spectacular elements of the built environment in cities – waterfront redevelopments, 'festival' settings, theme parks, signature buildings and so on – there are also more mundane elements 'that embody contemporary society's basic values and serve to enculturate individuals into its way of life' and so become defining phenomena of their era. Kellner (2005) calls these phenomena megaspectacles (though metaspectacles may be a more appropriate term) and cites McDonaldization as an example. For both Kellner and Debord, spectacle is a tool of pacification and depoliticization; a pervasive condition that stupefies social subjects. In submissively consuming spectacles, people are estranged from actively producing their lives.

The idea of neotraditional master-planned communities as a contemporary suburban 'metaspectacle' is consistent with this (Knox 2008). Janet Abu-Lughod (1992) has warned of the use and abuse of the notion of 'tradition' to reinforce or maintain established forms of domination. In other words, the idea of tradition is a rhetorical device, and 'traditions' themselves can be seen as being identified, manufactured, packaged and deployed in pursuit of social inclusion and/or exclusion. From this perspective, neotraditional architecture and traditional neighbourhood development are inherently socially regressive. Richard Sennett (1997: 67), for example, describes them as 'exercises in withdrawal from a complex world, deploying self-consciously "traditional" architecture that bespeaks a mythic communal coherence and shared identity in the past'. He concludes: 'Place making based on exclusion, sameness, and nostalgia is socially poisonous and psychologically useless: a self weighted with its insufficiencies cannot lift the burden by retreat into fantasy' (Sennett 1997: 69).

Lloyd and Clark (2001) have alluded to the contemporary metropolis as an 'entertainment machine', while Allen Scott (2001) describes significant portions of the contemporary metropolis as an 'ecology of commodified symbolic production and consumption':

> a place in which selected spaces are given over to ingestion of the urban spectacle, upscale shopping experiences, entertainment and night-time scenes, supplemented by occasional doses of cultural stiffening supplied by museums, art galleries, concert halls and so on. These spaces dovetail smoothly in both formal and functional terms with the gentrified residential neighborhoods and high-design workplaces that are the privileged preserve of the upper tier of the urban labor force in the modern cognitive-cultural economy.
>
> (Scott 2001: 17)

Among the emblematic elements of this ecology of commodified symbolic production and consumption are urban nightlife, 'starchitecture' and 'museumization'.

Commodification of urban nightlife

For a significant fraction of consumers – relatively young and relatively affluent – the principal component of the experience economy revolves around the urban night-time economy of clubs, discos, bars, pubs, restaurants and theatres. The expansion of the urban night-time economy has created new uses for city centres, transformed the design of cities and their transport networks, and altered meanings of public space, participation and citizenship (Hobbs *et al.* 2000). Abandoned industrial architecture, obsolescent cinemas and former banking halls have been sought out by corporate entertainment chains and transformed into clubs and bars.

Meanwhile, the emergence of new class fractions within the 'new economy', together with the growth of business tourism and corporate hospitality, has stimulated demand for stylized, safe and sanitized nightlife. At the same time, many cities, impelled by the neoliberal political economy of the past several decades, have pursued entrepreneurial urban policies that have sought to facilitate and promote the growth of their local night-time economies. This has taken the form of deregulation (with regard to alcohol and entertainment licensing) as well as public–private partnerships predicated on a profitable night-time component. As a result, many city centres previously considered marginal or even dangerous at night have been recast and are now promoted as being central to the image of a progressive, cosmopolitan city (Figure 6.1).

This expansion of the urban night-time economy has also created a new 'field', in Bourdieu's terms, for the deployment of both economic and cultural capital. Different class fractions and income groups represent niche markets within the urban night-time economy that corporate chains have exploited with carefully designed settings and branded chains of clubs, pubs and bars. In addition to stratification and segmentation in terms of price and exclusivity, the commodification of urban nightlife is, as Hollands and Chatterton (2003) observe, highly design-led. Different concepts and market segments call for different settings. There are 'chameleon bars' that function as comfortable, pub-like environments for shoppers and people wanting lunch or a drink after work but as nightclubs – with the music turned up and the lights turned down to attract a younger crowd – by night. There are exclusive and stylish café-bars that cater to local gentrifiers and to Bourdieu's 'new petit bourgeoisie' of well-paid junior commercial executives, medical personnel, editors, radio and TV producers and presenters, magazine journalists and the like. Other settings are designed to attract previously underrepresented groups such as women, gay and ethnic populations into commercial nightlife in packaged and sanitized formats such as gay villages, female-friendly bars and

Figure 6.1 The urban night-time economy: Grafton Street, Dublin. (Photo: Werner Dieterich/Getty Images)

ethnic entertainment zones. For the more traditional clientele the strategy is a combination of competitive pricing and novel experience based on scale and spectacle: 'superpubs' that can handle big crowds, disco-bars that feature an undercurrent of heightened sexuality:

> Scantily clad bar staff, striptease artistes, organized drinking games, hen nights, stag nights, and special nights for nurses and students are laid on, as well as fifty pence a pint nights, and the inevitable three shots of whatever's not selling well for a pound.
>
> (Hobbs *et al.* 2000: 707)

The overall result is a gentrification, homogenization and commodification that tends to displace traditional pubs and bars and to suffocate alternative and creative local development. Hollands and Chatterton (2003) point to the numerous examples from the United Kingdom:

> Chorion's *Tiger Tiger* brand, which started in London, with plans to roll out further premises in Leeds, Manchester and Birmingham, is described as an 'upmarket nightclub' or themed 'super club' (comprising bars, restaurant, club in one) catering for a 25 plus age group. The Luminar Group has talked of upgrading and stylizing many of the traditional night-clubs they inherited from Rank, and Surrey Free Inns (SFI) bought the *Slug and Lettuce* chain, described as 'yuppie' bars 'aimed at affluent urban professionals, offering quality food and premium beers', to complement its other two slightly less up-market brands *Bar Med* and *Litten Tree*. It is also worth noting that SFI own *For Your Eyes Only* venues, a chain of lap dancing bars aimed at 'corporate clients', which signifies a growing normalization of the sex industry within urban nightlife.
>
> (Hollands and Chatterton 2003: 377–378)

Globalization and 'starchitecture'

One of the consequences of capitalist globalization has been a transformation of the structural composition of architectural practice. Following an increasingly international clientele, more and more firms have developed a global portfolio of design work. Some of them are transnational corporations in their own right, huge architecture and engineering ('A&E') firms that have grown from what Robert Gutman (1988), in his pioneering study of the sociology of architecture, called 'strong delivery firms', commercial firms that rarely win awards but build a great deal. Others have grown from what he called 'strong-service firms', practices that are design-oriented but business-centred. A third group of global practices consists of what Gutman (1988) called 'strong-idea firms'. These are the global brand names of contemporary architecture, the 'starchitects' who are known for their signature buildings around the world (Knox 2010).

Meanwhile, the growth of the state fraction of the transnational capitalist class, as noted in Chapter 2, has intensified the importance of 'starchitecture', as cities compete for global status through promotion of signature buildings and the affect of celebrity. By choosing modern, high-tech and futuristic architectural designs from celebrity architects, local political and economic elites are able to create transnational urban spaces that cater for the needs of the transnational capitalist class. The symbolic capital of architectural design is transformed into other forms of capital in the process (Ford 2008; Ren 2008). And the growth of the consumerist fraction of the transnational capitalist class has supported the connections between consumption and design, including the marketability of the work of brand-name designers (Sklair 2005, 2006).

Donald McNeill (2009) observes that 'signature' architects like Zaha Hadid, Rem Koolhaas, Renzo Piano, Richard Meier, Norman Foster, Richard Rogers, Mario Botta, Frank Gehry, Santiago Calatrava, Jacques Herzog and Pierre de Meuron share certain characteristics. The first of these is an identifiable persona, whether through book authorship, television appearances, or simply personal style: just looking the part. But with many run-of-the-mill design professionals keen to demonstrate a personal sense of style, signature architects have to go a little further in order to stand out. McNeill (2009) quotes Adam Mornement's satirical characterization of different 'breeds' of global architect, including the

> thirty to forty-five-year-old 'visionary', a manifesto-writing, loft-dwelling centre-leftie wearing miniature glasses and 'anything tight-fitting and black'; and the 'contender' with a practice name of 'Me and Partners', 'a pragmatic balance between promoting the name of the dominant design partner, and acknowledging the contribution/presence of colleagues', who appears '50% academic/50% business guru – black linen suit (no tie), gracefully greying hair and a subscription to *Fortune*'.
>
> (McNeill 2009: 64)

A second common characteristic is an ability to develop a distinctive oeuvre or 'look', whether based on striking shapes, surfaces or concepts, that have an immediate, populist impact. The third common characteristic is a flair for self-promotion. In this, signature architects are aided and abetted by trade magazines, the architectural press, and the nexus of critics and editors whose profits and livelihood depend, in part, on sustaining the international star system. Celebrity is also propagated through major professional awards like the Pritzker Prize, while the pervasive emphasis on the cult of the individual in architectural education, with its almost unquestioned reverence for big names and emphasis on great exemplars and heroic architecture, also plays to the advantage of contemporary signature architects with a global brand.

Eventually, stardom becomes self-reinforcing, as real estate developers realize that brand-name architects can add value to their projects, city leaders compete to acquire the services of the top names to design signature buildings that will put their city on the map, and the signature buildings of star architects provide the backdrop for fashion shoots, movie scenes, TV commercials, music videos and satellite news broadcasts.

The Bilbao effect

The ability of a high-profile building of radical design to put a city on the global map was demonstrated by Sydney Opera House (Figure 6.2), designed by Danish architect Jørn Utzon in the late 1950s and completed in 1973. Utzon was not a star architect – indeed, he was relatively unknown and the Opera House was his first project outside Denmark and Sweden. But the structure, initially very controversial, is now recognized by UNESCO as a World Heritage Site and has become iconic of both Sydney and, indeed, of Australia as a whole. In spite of massive cost overruns in the construction of the building, the return on investment for Australia's government has been extraordinary. With the globalization of the economy, the increasing aestheticization of consumer sensibilities, the emergence of 'hard branding' in urban regeneration schemes, and the increasing sensitivity of city governments to the ways in which their cities are perceived by businesses and

Figure 6.2 Sydney Opera House. (Photo: author)

tourists, new signature buildings have become a common aspect of urban development (Jencks 2005, 2006; Sklair 2006).

In the commodified and celebrity-oriented global culture that is now pervasive, about all it takes for 'signature' status is for a building to be the product of a brand-name architect. When the building is also spectacular and/or radical in design, it can rebrand an entire city and elevate its perceived status within the global economy. This is what happened with Frank Gehry's Guggenheim Museum in Bilbao (Figure 6.3) and its success has prompted many other cities to engage brand-name architects in attempts to replicate the 'Bilbao effect'. As Dejan Sudjic observes:

> Sometimes it seems as if there are just thirty architects in the world. . . . Taken together they make up the group that provides the names that come up again and again when another sadly deluded city finds itself labouring under the mistaken impression that it is going to trump the Bilbao Guggenheim with an art gallery that looks like a train crash, or a flying saucer, or a hotel in the form of a twenty-storey high meteorite.
>
> (Sudjic 2005: 296)

The Bilbao Guggenheim is part of Bilbao City Council's ambitious revitalization process whose ultimate aim is to turn Bilbao into a flourishing international hub of culture, tourism and advanced business services. The city's strategy has been

Figure 6.3 The Guggenheim Museum, Bilbao. (Photo: Bob Krist/Corbis)

based around physical regeneration, featuring signature structures as symbols of modernity and an affect of economic revitalization (McNeill 2000). The Guggenheim Museum is part of the Abandoibarra riverside redevelopment scheme, the master plan for which was devised by César Pelli, Diana Balmori and Eugenio Aguinaga. Other notable developments include a 35-storey office tower (César Pelli), the Euskalduna Juaregia conference centre and concert hall (Federico Soriano and Dolores Palacios), the Bilbao International Exhibition Centre (César Azcárate), a new metro system with striking fan-shaped entrances (Norman Foster), a new airport (Santiago Calatrava), a footbridge spanning Nervión River (also Calatrava) and the 'Gateway' project, a mixed-use quayside development containing luxury flats, cinemas and restaurants (Arata Isozaki).

All this has been financed with support from the European Union, the Spanish and Basque governments and public–private partnerships. The Guggenheim itself cost US$110 million. It opened in 1997 and drew 1.36 million visitors in its first year, generating US$160 million in revenue. The numbers began to decline after three years but by then Bilbao had already become one of the leading weekend tourist destinations in Europe and the building itself had been featured in countless books, magazine articles, photo shoots and movie scenes, including the opening sequence of the 1999 James Bond film *The World Is Not Enough*. This success has become widely referred to as the 'Bilbao Effect' (or, alternatively the 'Guggenheim Effect'). As Anna Klingmann (2007: 7) observes, 'The building compounds use value, sign exchange value, and transformational value, converting the building into a piece of brand equity'. The image of the Bilbao Guggenheim simultaneously marketed several distinct institutions with great success: the Basque regional government, the city of Bilbao, the Guggenheim Foundation, and the architect, Frank O. Gehry and Associates Inc.

Not surprisingly, many other cities have sought to replicate the Bilbao Effect. Support for the construction of Oslo's new Opera House, for example, was framed around the Bilbao Effect (see Case Study 6.1). Often, however, the strategy has been limited to just one or two hopeful would-be 'signature' buildings, with correspondingly disappointing results. In Roanoke, Virginia, for example, a US$66 million 'Gehry-style' art museum (Figure 6.4) is the centrepiece of a downtown revitalization programme. Randall Stout, the architect, worked in Frank Gehry's office before forming his own practice in Los Angeles. He is described on his firm's own website as 'a visionary whose evocative design aesthetic consistently challenges architectural conventions'. But the Roanoke museum, post-Gehry, does not seem particularly radical or visionary. Stranded in an enervated downtown, the museum prompts an affect of embarrassment: a trying-too-hard building that ends up as a rather lame gesture of ambition.

Figure 6.4   Taubman Museum of Art, Roanoke. (Photo: author)

Case Study 6.1   **Oslo's Opera House**

Opened in 2008, the Norwegian National Opera and Ballet in Oslo (Figure 6.5) is part of a waterfront regeneration scheme branded as Fjord City. Oslo itself is situated in a natural amphitheatre at the northern end of Oslo fjord, with the plaza of the City Hall fronting the fjord. For the most part, though, the city had developed in a way that resulted in it turning its back to the sea, leaving the fjord coast to the docks, warehouses and heavy industries that eventually became obsolescent. In the 1980s, the western docklands of the Aker Brygge district were redeveloped, converting the city's last remaining dockside into dwellings, shops and office space, including the corporate headquarters of several of Norway's largest companies. It was intended to be the first step in a more extensive process of regeneration of the entire industrialized shore area but an economic slump halted the plans, leaving the run-down Bjørvika district on the eastern part of the shore isolated from the rest of the city by the busy E18 motorway.

The Fjord City project includes the rerouting of the E18 through an underwater tunnel, thus opening up the Bjørvika district to the rest of the city. The master plan for the site includes twelve medium-rise buildings incorporating up to 5,000 apartments, together with new buildings for a relocated Munch Museum, the Stenersen Museum of contemporary art, and the Oslo Public Library. The centrepiece of the

Figure 6.5 Norwegian National Opera and Ballet, Oslo. Designed by the architectural firm Snøhetta, the Opera House won the European Union Prize for Contemporary Architecture, the Mies van der Rohe Award for 2009. (Photo: Royal Press Nieboer/Corbis)

site, right on the fjord, is the Opera House, designed by the Norwegian firm Snøhetta. The competition brief for the Opera House stated that it should be of high architectural quality and should be monumental in its expression. It is one of three EU Eco-Culture projects that focus on energy efficiency in cultural buildings.

The construction and siting of the new opera house was hotly contested for years. The Norwegian National Opera, the national government and Norway's Conservative Party wanted to locate the new Opera House in the vicinity of the City Hall, while politicians from the eastern districts of the city lobbied for the Bjørvika site. As Marius Hofseth (2008) observes:

> The case for Bjørvika gained momentum as the potential for urban development became clearer. At the outset, the proponents had largely focused on the potential for urban renewal in economically and socially challenged areas of the inner eastern districts. Gradually, however, attention shifted towards the potential gains for the city as a whole. Following the opening of the Guggenheim museum in Bilbao in 1997 and its widely acclaimed

success as a stimulant for urban renewal, the argument about a 'Bilbao effect' in Oslo was widely cited. The city of Oslo hosted a 'Bilbao conference' in the city hall to develop these ideas further.

(Hofseth 2008: 102)

Key reading

Marius Hofseth (2008) 'The New Opera House in Oslo – A Boost for Urban Development?', *Urban Research and Practice*, 1.1: 101–103.

The Bilbao Effect itself has not been without criticism. One common charge is that the strategy of upscale physical revitalization has failed to attract international capital and advanced business services to Bilbao: something that was initially anticipated to be a consequence of the lure of spectacle and international-quality cultural amenities. Another disappointment, then, of design determinism. A further criticism of the strategy of upscale physical revitalization is that it ignores and even compounds social issues and the interests of lower-income households. Similarly, while the strategy provides prestigious cultural showcases, it contributes little or nothing to local cultural production while drawing public funds away from other cultural activities (Miles 2005). Finally, the success of the strategy in improving the quality of the city's environment has prompted a secondary 'Guggenheim Effect', inducing a certain amount of gentrification and speculative commercial development that has driven up land and housing prices and restricted the housing opportunities of younger and less affluent households (Vicario and Monje 2003).

Museumization and the experience economy

One of the most prized commissions among 'strong-idea' firms and would-be starchitects is a new museum. This is largely because museums offer an opportunity to design a sculptural building with a high-quality finish. Museums also pose important challenges in terms of the quality of lighting, the feel of public spaces, and the functionality of back-of-house spaces. The cultural importance of museums, meanwhile, guarantees a much greater level of publicity than, say, an office building or a courthouse. Most established starchitects are known for one or more major museum commission. Prominent examples include Frank Lloyd Wright (the original Guggenheim, New York), Richard Rogers (Centre Pompidou),

Figure 6.6 Quadracci Pavilion, Milwaukee Art Museum, designed by Santiago Calatrava and completed in 2001. (Photo: Don Klumpp/Getty Images)

Frank Gehry (Bilbao Guggenheim; Vitra Design Museum), Santiago Calatrava (Milwaukee Art Museum – Figure 6.6), Rem Koolhaas (Seoul National University Museum of Art), Daniel Libeskind (Jewish Museum, Berlin; Imperial War Museum, Manchester), Richard Meier (Getty Center, Santa Monica), Norman Foster (Museum of Fine Arts, Boston; Sainsbury Centre for Visual Arts, Norwich), Renzo Piano (Centre Pompidou, Paris; Whitney Museum of American Art, New York), Mario Botta (MoMa, San Francisco), and Herzog and De Meuron (Tate Modern, London).

Fortunately for aspiring starchitects, the growth of the experience economy has meant that there has been a boom in museum building in the past several decades. In the United States alone, over 600 new museum buildings were constructed in the 1980s and 1990s, and it is estimated that half of the art museums in the United States have been opened since 1970 (Evans 2003). The museum boom was set off in the mid-1970s by the completion of Centre Pompidou in Paris (Figure 4.3), as much a cultural amusement park and culture café as a museum. The boom was consolidated by the success of the redeveloped Louvre, with its completely rethought entrance halls and shopping corridors beneath I.M. Pei's distinctive and immediately recognizable glass pyramid in the central court of the ancient building (Figure 6.7).

As Elizabeth Wilson (1995) observes, the new Louvre

> looks more like an airport, or possibly a bank than an art gallery. The pyramid itself is exciting, but the open escalators, the shiny marble and the long row of shops, all dedicated to marketing various kinds of Louvre artefacts, speak corporate culture rather than aesthetic pleasure.
>
> (Wilson 1995: 158)

Here is the clue to the museumization of urban landscapes: the capacity of the contemporary museum to combine spectacle with consumption. Since the mid-1980s, the museum has been transformed 'from a place for the conservation of culture to a tourist destination where consumption is the primary mode of behaviour' (van Aalst and Boogaarts 2002: 198). Elizabeth Wilson again:

> The museum . . . has been transformed from educational instrument to something approaching, in some cases, a branch of Liberty's. . . . Today the trip to the museum – or the art exhibition – is quite definitely organized to lead up to the final stop of the shop, where, in buying coloured postcards of your favourite paintings (to be sent to equally aesthetically aware friends on suitable occasions) you set the seal on your own good taste. The museum café-restaurant, too, is by now a well established and lucrative branch of the catering industry (remember the Victoria and Albert Museum's 1980s ad – 'An ace caff with a rather good museum attached').
>
> (Wilson 1995: 156)

Figure 6.7 The Louvre pyramid. Commissioned by then French President François Mitterrand in 1984, designed by I.M. Pei. (Photo: author)

Thus museums become 'cultural supermarkets', with museum stores selling all kinds of designer products and souvenirs:

> It is not unusual for museums to actively exploit the commercial opportunity created by big retrospective exhibits, which draw huge crowds. These shows are recognized as merchandising opportunities for all sorts of souvenirs – Van Gogh sweatshirts, Mondriaan appointment books, Mendini pastry forks, and so on.
>
> (van Aalst and Boogaarts 2002: 197)

The biggest shop in the world for art books is in the Tate Modern in London, while the Museum of Modern Art in New York can boast higher retail sales per square foot than an out-and-out retailer like Wal-Mart.

As museums have proliferated, competition among them has intensified the importance of the experiential aspects of consumption. Visitors expect spectacle. As a result, the traditional core activities of large museums – the conservation and restoration of a permanent collection and the pursuit of scholarly research – have been displaced by the need to offer blockbuster themed shows that provide ticket and merchandise revenue. Museums are increasingly becoming temporary exhibition spaces for travelling exhibits.

Many of the new museums are specialized, catering to niche markets and special interests: sports, music, transportation and so on. Given the aestheticization of everyday life, it is not surprising that specialized *design* museums have proliferated. Examples include the Triennale di Milano, the Design Museum in London (Figure 6.8), the Cooper-Hewitt Design Museums in New York and Washington, DC, the Designmuseum at the University of California at Davis, the Vitra Design Museum in Weil am Rhein, the A+D Museum, Los Angeles, the Wolfsonian in Miami Beach, the Museum of Design in Atlanta, Red Dot Design Museum in Essen, and the Designmuseo, Helsinki. Such museums play an increasingly influential role in the construction of popular understandings of design, reinforcing the nexus of architecture, fashion and consumption.

The location of museums within cities is critical to their success. Graeme Evans (2003) has noted the association between location of museums and galleries and major city parks, suggesting that it links their dual recreational nature and also reinforces the privileged zones of the city:

> These central parks are often located in the well-heeled residential, office and shopping districts, and placing major museums there also enhances their symbolic and land value. Examples include the Mall area in Washington, DC; Rotterdam's museum park; Mexico City's Chaputalpec Park which contains over eight museums; São Paolo's museums of Modern Art, Folklore and Aeronautics in parque do Ibrapuera; Barcelona's Montjuic Park housing the Ethnological and Archaeological museums, the Art Museum of Catalonia and

Figure 6.8   The Design Museum, London, opened in 1989 in a renovated Thameside warehouse near Tower Bridge. (Photo: author)

Joan Miró museums; the Cité des Sciences et de l'Industrie in Parc de La Villette, Paris and the Burrell and Kelvingrove in Glasgow.

(Evans 2003: 430)

In many cities, clusters of museums, often with interrelated themes, have become a key element of the tourism sector and an important contributor to the urban economy. One example is the Museums Quartier in Vienna (de Franz 2005), a museum complex on the site of former imperial stables. Now one of the ten largest cultural complexes in the world, the Museums Quartier includes the Kunsthalle Wien, the Leopold Museum (with its collections of modern Austrian art) and the Museum of Modern Art Ludwig Foundation Vienna (which houses one of the largest European collections of modern and contemporary art). In Portsmouth, England, the theme of the city's cluster of museums is nautical, and includes the Royal Navy Submarine Museum, the Royal Marines Museum, the Explosion Museum of Naval Firepower and the D-Day Museum, as well as the Historic Dockyard museum that features the *Mary Rose*, HMS *Victory* and HMS *Warrior* (see also Case Study 5.2).

In Amsterdam, the Rijksmuseum, Van Gogh Museum, Stedelijk Museum and Diamond Museum are clustered around the Museumplein ('Museum Square'). In Berlin, the Altes Museum, Neues Museum, Alte Nationalgalerie, Bode Museum and Pergamon Museum are all located on Museumsinsel ('Museum Island') bounded by the River Spree. The most celebrated cluster of museums is perhaps along 'museum mile', on Fifth Avenue in New York between 82nd and 105th

Streets. It includes the Museum of the City of New York, the Jewish Museum, El Museo del Barrio, the Cooper Hewitt National Design Museum of the Smithsonian Institution, the National Academy Museum and School of Fine Arts, the Solomon R. Guggenheim Museum, the Neue Galerie, the Goethe-Institut and the Metropolitan Museum of Art. Each year, these nine museums collectively open their doors for a day for a mile-long, traffic-free block party and arts festival that attracts over 50,000 visitors.

Cities of culture and design

Museumization is only one dimension of the rebranding of cities as centres of culture and design. Another – often closely related – is the sponsorship of arts festivals and cultural events. One of the best known examples is the Edinburgh Festival, which has become a key element in the image of the city, as well as a big generator of revenues from tourists. As Bernadette Quinn (2005) observes, festivals

> have taken on a new significance in the context of globalization. They are now construed as entrepreneurial displays, as image creators capable of attracting increasingly mobile capital, people and services. Major events are seen as being particularly effective in that they ally tourism objectives with urban planning, while simultaneously providing a means through which political and urban élites can refashion collective feelings of identity, emotion and consciousness.
>
> (Quinn 2005: 931)

To compete successfully at the national and international level, however, cities with aspirations of managing their brand image as a centre of culture and design must deploy a sustained and multilayered strategy.

Brand management, design and regeneration

Barcelona is often cited as the exemplary case of urban regeneration and brand management through the promotion of culture and design (Casellas and Pallares-Barbera 2005). Systematically peripheralized under the Franco regime, the city's renaissance began with its hosting of the Olympic Games in 1992, which provided the political impetus to revive Catalan identity. Place promotion, tourism campaigns, and television and magazine advertising all utilized images of the city's legacy of architecture, plazas and seafront bars and restaurants. Significant investments were made in improving this physical infrastructure, while brand-name architects were commissioned for new flagship buildings and urban design projects. Incrementally, there was a 'wholesale ideological, cultural, commercial and hence aesthetic repositioning' of the city (Julier 2000: 126).

Gradually, Barcelona acquired an image as a place where residents and tourists alike could indulge taste and lifestyle with an emphasis on culture and the arts. Equally important, the affect associated with the city changed significantly for the citizens of the city themselves, as Guy Julier (2000) notes:

> New design in Barcelona infiltrated all levels of everyday life by the late 1980s. The values of modernity, Europeaneity and technology but also locality and creativity were invoked throughout the cityscape. Its significance was under-lined by a series of interconnected ideological and cultural features: it contrasted starkly with Spain's immediate past by replacing national isolationism, conservatism and environmental chaos with internationalism, modernity, and civic pride; it foregrounded new design as an amenity for a new civil society; it aspired to achieving a confluence between local identity, regional political will and modernity in a transnational context. Design hardware – the city's new bars, public spaces, street graphics – was deployed in an intensely public arena both in itself but also in its constant mediation and reproduction through the media. It served to bolster the emotional software of the city, promoting attitudinal aspirations and unity by linking nationalistic sensibilities, self-identity and place. The consistency of its message in terms of taste and cultural capital gave the appearance of local cohesion.
>
> (Julier 2000: 128)

The European Union has sponsored urban brand management and the promotion of culture and design since the mid-1980s as a way of promoting both regional development and the notion of a common European cultural heritage. Its Capitals of Culture programme was conceived in 1983 by Melina Mercouri, then Greek Minister for Culture. Athens was duly appointed the first European Capital of Culture in 1985. Subsequent holders of the title include Florence, Amsterdam, West Berlin, Paris, Glasgow, Copenhagen, Thessalonica, Stockholm, Prague, Krakow, Porto, Rotterdam, Bruges, Graz, Genoa, Helsinki, Lille, Cork, Patras, Sibiu and, most recently, Stavanger, Liverpool (Figure 6.9), Vilnius and Linz.

Financial support from the EU for these cities amounts to only a few hundred thousand euros in each case, but the real value is in branding and recognition. Glasgow is widely acknowledged to have made the most of the programme (see Case Study 6.2). Meanwhile, UNESCO has got in on the act, designating Kobe, Nagoya, Berlin, Montréal and Buenos Aires as Cities of Design, part of UNESCO's Creative Cities Network under the framework of its Global Alliance for Cultural Diversity.

Such programmes, observes Graeme Evans (2003: 427), have created 'a growing tier of peripheral and regional cities that repeatedly enter such competitions and justify major public investment in new venues and transport in terms of the regenerative benefits of branding that will accrue'.

Figure 6.9 Liverpool, the European Union Capital of Culture in 2008. (Photo: Christopher Furlong/Getty Images)

Case Study 6.2 **Glasgow's rebranding exercises**

One of the earliest and by some measures most successful attempts to rebrand a city was in Glasgow. In the mid-1980s Glasgow was a prime example of industrial decline. Heavy job losses, mostly in the manufacturing sector, meant that total employment fell from almost 560,000 in 1950 to less than 400,000 in the mid-1980s. The city's reputation was one of social polarization, a city well past its prime; full of character, but riven by sectarian tensions and bowed by poverty and unemployment. In 1983 the city improbably launched its *Glasgow's Miles Better* advertising campaign to promote itself as a tourist destination and a location for industry. The innovative feature of the campaign was the use of the figure of smiley-face Mr Happy from the Mr Men children's book series authored by Roger Hargreaves. Soon afterwards, the city sought recognition as a European Capital of Culture as a further step in overhauling its image. The designation was eventually granted in 1990. Important buildings were duly cleaned and floodlit, drawing attention to the built environment, one of the city's great assets. In 1999 Glasgow was recognized with the prestigious British City of Architecture and Design designation.

Meanwhile, inspired by Barcelona's success, Glasgow began to take its rebranding and regeneration seriously, developing both policies and infrastructure relevant to the culture of design. The Lighthouse, Scotland's Centre for Architecture, Design and the City, was opened in 1999 after a £13 million conversion of Charles Rennie Mackintosh's derelict 1895 Glasgow Herald newspaper office. It is now a city-centre flagship building, with gallery space, a rooftop café and education facilities. In 2004 a new slogan was launched: *Glasgow: Scotland with Style*. Building directly on the success of the original *Glasgow's Miles Better* campaign and the Capital of Culture billing, the intention was to increase tourism, investment and civic pride in the city. Funded by Glasgow City Council, the European Regional Development Fund and the Greater Glasgow & Clyde Valley Tourist Board, the Scotland With Style campaign has featured the BBC Proms in the Park at Glasgow; the sponsorship of a clipper boat named *Glasgow: Scotland With Style* in Round-the-World Yacht Races; and the sponsorship of the Glasgow: Scotland With Style Design Collective, which showcases the work of emerging Glaswegian fashion designers at London Fashion Week. Overall, Glasgow has been successful through successive branding campaigns in positioning itself as a centre of creativity and design, its image and reputation much improved – even though economic benefits such as job creation seem questionable and unemployment and poverty in the city remain above average.

Key readings

Beatriz García (2005) 'Deconstructing the City of Culture: The Long-term Cultural Legacies of Glasgow 1990', *Urban Studies*, 42.5–6: 841–868.

A. Jones, N. Lee, L. Williams, N. Clayton and K. Morris (2008) 'How Can Cities Thrive in the Changing Economy?', *Ideopolis II Final Report*. London: The Work Foundation.

In winning cities, regeneration often results in rather formulaic outcomes. Evans (2003) describes what happened in Helsinki, one of the EU's Capitals of Culture for 2000, as follows:

> The city has embraced the cultural industries and arts flagship strategy, which includes developing a cultural production zone centred on the Cable Factory and music conservatoire, and a cultural consumption quarter around the Glass Media Palace. This 'Palace' – a row of shopfronts and cafés backed by a central coach station – was an inheritance from the ill-fated 1930s Olympic Games, converted to an art house cinema, art book shops, cafés and media production facilities to form part of a cultural triangle with the new museum of

contemporary art, a multiplex cinema and a planned tennis palace museum. The obligatory contemporary art museum, Kiasma ('a crossing or exchange'), designed by American architect Steven Holl and, like Guggenheim-Bilbao, hosting a private, imported collection, 'seeks to redefine the art museum as an institution, shifting from the image of elitist treasure house to public meeting place'. This new curved asymmetric building includes a ground floor café which spills out into the outdoor space, with its fountain, lawn and hardscape, a formula now familiar in the design of the modern office and shopping mall.

(Evans 2003: 426)

Nevertheless – and in the absence of other ideas – strategies of urban regeneration based on the promotion of culture and design remain attractive to the leadership of many cities. In many cases, it is tied to Richard Florida's prescriptive ideas about how to attract and retain a 'creative class' (see Chapter 2) that is held to be pivotal to the New Economy/Dream Economy/Experience Economy. According to Florida (2002: 13), this means that if cities do not make strong efforts to establish the right 'people climate' for creative workers, they will 'wither and die'. It means, in addition to promoting tolerant, diverse and open communities, having 'authentic' historical buildings, converted lofts, walkable streets, plenty of coffee shops, street art and a street culture with experiential intensity. An affect, in short, that is 'seething with the interplay of cultures and ideas; a place where outsiders can quickly become insiders' (Florida 2002: 227).

There are, of course, many different approaches to urban entrepreneurialism based on the promotion of culture and design. Some cities have focused on 'high' culture and the provision of museums, galleries, theatres and concert halls; others emphasize tradition, history and monuments; others still focus on their package of nightlife – theatres, concert halls, nightclubs, red-light districts, cafés and restaurants. Some extend their efforts to social inclusion and liveability, using urban design to facilitate community interaction, to enhance sociability in relation to housing and public space (such as streets and parks), to contribute to crime prevention and to improved transport infrastructure (Evans 2001). Graeme Evans (2005) has identified three broad approaches through which cultural activity is incorporated (or incorporates itself) into the urban regeneration process:

- *Culture-led regeneration*: This approach typically involves a high-profile cultural flagship or complex that serves to symbolize the city's commitment to cultural activity. It may involve the design and construction (or reuse) of a building or buildings for public or mixed use; the reclamation of open space (for example, garden festivals, EXPO sites); or the introduction of a programme of arts 'festivals', events and public art schemes. The case of Helsinki, described above, is an example of this approach.

- *Cultural regeneration*: In this approach, cultural activity is more integrated into a spatial strategy alongside environmental, social and economic activities. Barcelona is the exemplar.
- *Culture and regeneration*: In this 'default' approach, cultural activity is not fully integrated in strategic planning, often because the responsibilities for cultural provision and for regeneration sit within different city departments. Cultural interventions are often small: a public art programme for office development, once the buildings have been designed; a heritage interpretation programme or local history museum.

It is increasingly the case that particular areas within cities are branded as creative districts or cultural quarters: El Raval in Barcelona, the Northern Quarter in Manchester, and the Zona Tortona in Milan, for example. Matthew Wansborough and Andrea Mageean (2000) identify five key characteristics of such quarters:

(1) Central location within the city, frequently adjacent to major retail or commercial areas. The central location makes such areas more accessible and also invites less formal usage, which is in character with many of the activities that occur here (e.g. bars and cafés), as well as making them ideal centres for specific uses and specialist interests (e.g. small retailers and night-clubs, etc.).

(2) Cultural facilities concerned with both consumption and production, i.e. music venues and recording studios, cinemas and film schools, market stalls and craft workshops. However, most cultural quarters tend to become centres of consumption (i.e. tourist attractions) rather than providing a balance of the two. Less formal facilities are also required, such as the street and the square, in order to accommodate programmed events and festivals, etc.

(3) Mixed use allows for economic diversity, provides a more human-scale environment and helps to increase the sense of containment and self-sufficiency of the area. A mixture of small-to medium-scale businesses (shops, studios, and performance venues), cafés, bars, pubs, clubs, hotels, cinemas and theatres as well as residential developments allows for diversity and activity at all times.

(4) 'Cross-over' between production and consumption. Due to the relatively high value-added nature of the production process for many cultural industries, it is important that there are close (geographical) links between the point of production and the point of consumption. Due to the smaller scale and local mix, businesses are more able and willing to share or use each other's resources, skills and facilities, etc. It is this general relationship between cultural consumption and cultural production that is a crucial factor for both the general functioning and the successful functioning of cultural quarters.

(5) Public art and its integration with the built environment. Once again, this calls for a balance between production and consumption, as local artists can be used to create attractions for their local environment. It also suggests that cultural quarters should be characterized by good urban design and, consequently, by a vital and vibrant public realm. This is achieved by creating art that engages

and involves people with the environment in order to contribute to a greater understanding of the area.

(Wansborough and Mageean 2000: 188)

All of these approaches are vulnerable to the criticism that they are not as beneficent as they may seem. As Jamie Peck (2005) points out, there is, at minimum, the risk that flagship arts facilities, urban design interventions, and the like can easily lapse into their own kind of generic landscapes – which may lead them to be spurned by the highly valued 'creatives' in their restless search for authentic experiences. Peck also notes how culture-based approaches to urban brand management and regeneration fit closely with the neoliberal agenda:

> The reality is that city leaders from San Diego to Baltimore, from Toronto to Albuquerque, are embracing creativity strategies not as *alternatives* to extant market-, consumption- and property-led development strategies, but as low-cost, feelgood *complements* to them. Creativity plans do not disrupt these established approaches to urban entrepreneurialism and consumption-oriented place promotion, they *extend* them. . . . Rather than 'civilizing' urban economic development by 'bringing in culture', creativity strategies do the opposite: they commodify the arts and cultural resources, even social tolerance itself, suturing them as putative economic assets to evolving regimes of urban competition.

> (Peck 2005: 761, 763; emphases in original)

Designer cities, like designer clothing, kitchenware and furniture, are geared almost exclusively toward affluent and consumption-oriented households. But unlike clothing, kitchenware and furniture, cities belong to a wider society. In this context, it is important to reiterate the point made at the beginning of Chapter 1: the fundamental importance of design to cities and urban life rests on its roles in supporting and sustaining the political economy of urbanized capitalism. Design has an unambiguous role in facilitating the circulation and accumulation of capital and the disguising, concealing and transforming powers of design are increasingly important in globalized, consumption- and experience-oriented economies.

Summary

All city spaces are shot through with complex layers of intersubjective meanings and affect. Design plays an increasing role in shaping and facilitating these feelings as the experience of distinctive places, physical settings and landscapes has become an important element of consumer culture. Globalization has meanwhile prompted communities in many parts of the world to become much more conscious of the ways in which they are perceived by tourists, businesses, media firms and consumers. As a result, places are increasingly being reinterpreted, reimagined, designed, packaged and marketed. This can be seen as part of the 'experience economy' that is intimately co-dependent with the 'romantic capitalism', 'self-illusory hedonism' and 'dream economy' of competitive consumption. The need for the serial enchantment and re-enchantment of products and places has intensified the significance of spectacle, simulation, theming and sheer size in contemporary material culture, and the importance of the affect of places is something that is increasingly recognized in professional contexts.

The overall result is that parts of cities become an ecology of commodified symbolic production and consumption. Emblematic elements of this include commodified urban nightlife, 'starchitecture' and 'museumization'. The emergence of new class fractions within the 'new economy', together with the growth of business tourism and corporate hospitality, has stimulated demand for stylized, safe and sanitized nightlife. Meanwhile, the growth of a transnational capitalist class has intensified the importance of 'starchitecture' as cities compete for global status through promotion of signature buildings and the affect of celebrity. The success of Frank Gehry's Bilbao Guggenheim, together with Bilbao's strategy of featuring signature structures as symbols of modernity and an affect of economic revitalization, has prompted a 'Bilbao effect', as other cities attempt to rebrand themselves and spark urban regeneration through the promotion of culture and design.

Further reading

Charles Jencks (2005) *The Iconic Building*. New York: Rizzoli. Jencks traces the intersection of politics, finance and the media in explaining the trend toward starchitects and starchitecture.

Anna Klingmann (2007) *Brandscapes: Architecture in the Experience Economy*. Cambridge, MA: MIT Press. Shows how branding and experience management are at the forefront of contemporary architectural theory and practice, and suggests that commodified desire may give designers more control over business plans and urban politics.

Donald McNeill (2009) *The Global Architect: Firms, Fame and Urban Form*. London: Routledge. Explores the influence of globalization processes on urban change, architectural practice and the built environment, emphasizing the 'star system' of international architects and the role of advanced information technology in expanding the geographical scope of the industry.

George Ritzer (2009) *Enchanting a Disenchanted World: Continuity and Change in the Cathedrals of Consumption*, 3rd edn. Thousand Oaks, CA: Pine Forge Press. Provides a rich array of examples in analysing new patterns and spaces of consumption.

7 Design services and the city

This chapter looks at big-city environments as crucibles of creativity and innovation in design. Globalization has increased the number of design firms with a global portfolio of work and global networks of branch offices. Because various design services and creative-products industries are constantly engaging one another, sharing ideas and resources, design services, global and local, tend to be localized in distinctive communities of practice. These are often reinforced by the sociality and buzz that is important to cultural production. The agglomeration of design services within cities is underpinned by localization economies and by 'fabric effects' related to the economic and aesthetic attributes of the built environment. Successful communities of design practice add to a city's overall status and can influence not only the city's brand image but also that of other products associated with the city.

Large cities are the natural setting for design services of every kind. In the United States, there were more than 595,000 designers in 2006, with each professional category of design expected to grow significantly by 2016 (Table 7.1). Without exception, these jobs are highly localized in major metropolitan areas. New York, Chicago, Los Angeles, Boston and San Francisco dominate, especially with regard to concentrations of architects; but Detroit and San José host prominent concentrations of industrial designers, while Seattle tops the list among graphic designers (Markusen and Schrock 2006). In the United Kingdom, about 185,000 people were employed in design in 2005, accounting for an estimated turnover of £11.6 billion. Almost half were located in London and metropolitan South East England, with secondary concentrations in Manchester, Leeds, Birmingham and Bristol (Design Council 2005). Elsewhere in Europe, there are major concentrations of designers in Barcelona, Berlin, Milan and Paris, with secondary concentrations in Amsterdam, Helsinki, Madrid, Prague, Rome and Vienna.

Table 7.1 Design employment, United States, 2006

	Total employment 2006	Percentage change 2006–2016
Architects	132,000	18
Landscape architects	28,000	16
Planners/urban designers	34,000	15
Graphic designers	261,000	10
Industrial designers	48,000	7
Interior designers	72,000	19
Fashion designers	20,000	5

Source: United States Bureau of Labor Statistics, *Occupational Outlook Handbook, 2008–09* (www.bls.gov/oco)

As we saw in Chapter 2, big-city environments are crucibles of creativity. They are forcing houses of cultural innovation and important arenas for the creation of taste. Most large cities contain specialized districts that act as 'creative fields', distinctive settings that are rich with innovative energy, dense interpersonal contacts and informal information exchange. Design can only really flourish, in fact, in settings where knowledgeable professionals are in close contact with one another, with clients, and with other creative individuals. Innovation in design results from the combination of a wide range of different types of knowledge, as designers synthesize and recombine ideas so as to produce novel effects and new designs. The same is true in other creative industries. Economist Richard Caves (2000) has noted that creative products industries are characterized by what he calls the 'motley crew' property: that they rely on combinations of diverse people, groups and industries.

As Elizabeth Currid (2007: 15) points out, Andy Warhol understood that the various branches of design and creativity are 'constantly engaging one another, sharing ideas and resources' and that 'there is an important social dimension to cultural production'. All of these elements were incorporated in Warhol's famous Factory in Manhattan. Currid uses the dynamism and hybridity of Warhol's Factory setting as a metaphor for the creative economy within New York City. Currid also notes that Warhol understood that, as well as transforming commodities into art (as in his famous paintings of Campbell's soup cans), art can be translated into commodities (not least in the posters, paperweights, magnets and tableware featuring Warhol's pop art). As Elizabeth Wilson (2006) points out, this aspect of creativity is central to the nature of urban modernity, in particular its double-sided character, characterized simultaneously not only by new forms of constraint and commodification in everyday life, but also by new possibilities for active experimentation and identity formation.

Communities of practice and the ecology of design

The interdependence of design with other creative activities, with the social scene in cities, and with the broader political economy of urbanized capitalism means that the design professions tend to exhibit high degrees of geographic concentration within cities, especially in great world cities such as London, Los Angeles, New York and Paris. Within broader urban ecologies that sustain neo-bohemias, art worlds, gentrifying neighbourhoods, branded cultural quarters and 'semiotic districts' (of specialized luxury retail stores, expensive restaurants, cafés, art galleries and antique shops) are distinctive communities of practice based on clusters of design professionals.

These clusters are important both to the professionals that live and/or work in them and to the economic dynamism of many cities. They are central to the 'creative fields' of Allen Scott's (2008) 'resurgent metropolis'. In major world cities, the net result goes beyond the point of just having concentrations of fashion designers, architects and so on to actually being a setting where influential cultural ideas and trends emerge, fueling and reinforcing economic and cultural globalization.

Agglomerative tendencies

Agglomeration is characteristic of a broad spectrum of activities within what has been variously described as the 'new economy' and the 'dream economy', and especially characteristic of the cultural-products industries that involve design. Such agglomerations constitute important features of the spatiality of the 'resurgent metropolis' described in Chapter 2. The fundamental reasons for such agglomeration were recognized long before the growth of cultural-products industries. Alfred Marshall famously noted in 1890 that many firms seek to benefit from what are now referred to as *external economies* (cost savings resulting from advantages that are derived from circumstances beyond a firm's own organization and methods) by locating in specialized spatial clusters of industry and human capital. As he put it, 'So great are the advantages which people following the same skilled trade get from near neighbourhood to one another. The mysteries of the trade become no mystery: but are, as it were, "in the air" ' (Marshall 1890: IV.X.6, § 3).

The dynamics and anatomy of clusters of cultural-products industries have attracted a good deal of attention from geographers and planners (Cinti 2008; Currid and Connolly 2008; Lorenzen and Frederiksen 2008). A first observation from such studies is that clusters of cultural-products industries exhibit both a horizontal and a vertical dimension (Bathelt *et al.* 2004). The *horizontal* dimension involves firms that produce similar services and compete with one another. These firms benefit from co-location because it allows them to be well informed about the characteristics of their competitors' businesses and about the quality and cost of their services. Proximity allows rivals to monitor and evaluate alternative practices and solutions. It results in 'knowledge spillovers' and flows of ideas that are recombined and synthesized to produce innovative designs and solutions. Proximity can also promote trust, shared conventions and standardized business practices, which help to make all of the firms in the cluster more efficient (Maskell and Malmberg 1999; Rantisi 2002a).

The *vertical* dimension of clusters consists of firms that are complementary, linked through a network of supplier, service and client relations. Design firms develop backward linkages to businesses that provide specialized services (visualization software, engineering consultancies, colour forecasting consultancies and photography, for example) and facilities (studios, galleries, exhibition spaces). Forward linkages develop as other firms of many kinds (from advertising to ready-to-wear clothing and real estate development) take advantage of design services. This interdependency of firms results in the sort of distinctive 'industrial atmosphere' recognized by Marshall.

As with other industries, the initial advantages enjoyed by successful local clusters of cultural-products firms are consolidated over time by 'localization economies' such as access to a local pool of labour with special skills and experience, access to specialized schools and research institutes, and access to marketing organizations and cultural gatekeepers. The cluster is deepened and extended as more firms are attracted and more and more skilled and aspiring workers are drawn in. This is 'cumulative causation', the self-reinforcing, spiralling build-up of advantages based on the combination of external economies and localization economies.

An analysis of 1,676 architecture and engineering offices in London found that about a third of them were highly clustered, with at least ten other A&E offices located within 100 metres of them (Taylor *et al.* 2003). This is a relatively high degree of clustering, but less than for banking, insurance and business support services, more than half of which are spatially clustered. Within London, there are two broad clusters of A&E offices (Figure 7.1). The first is in the West End, where banking, real estate, law, management consultancy, accountancy and advertising are also clustered among the high-end shopping and high-status residential neighbourhoods east of Hyde Park and south of Regent's Park. The second is along the northern fringe of the City of London, sandwiched between the financial core and the gentrifying neo-bohemias of Shoreditch and Hoxton, an emergent cluster of design and advanced business services that includes information technology and new media firms.

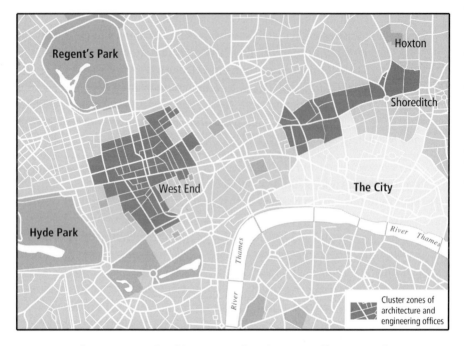

Figure 7.1 Cluster zones of architecture and engineering offices in London. (Source: After D.R.F Walker and P.J. Taylor (2003) *Atlas of Economic Clusters in London,* **www.lboro.ac.uk/gawc/visual/conatlas.html)**

The socio-spatial ecology of design

In the case of clusters of design services, these agglomerative tendencies and dynamics are reinforced by some important aspects of sociality. Perhaps the most straightforward of these is the importance for designers of personal interaction with clients. As sociologist Paul Kennedy (2005) observes:

> Deep sociality remains unavoidable and crucial while being rooted inevitably into particular locations even though these simultaneously function as global workplaces. Warehouses, ports, railway stations, factories, airports, law courts and the rooms where barristers and clients bargain, company boardrooms, the bars and offices that allow negotiations and deals to be discussed and agreed between insurance agents, bankers, business consortiums and consultants, the studios where advertising team struggle with clients to create themes that sell – all these and many more thrive as essential, vibrant locations where social actors must operate together in order to make global as well as national and regional economies work.
>
> (Kennedy 2005: 193)

One of the key attributes of cultural-products industries identified by economist Richard Caves (2000) is the 'nobody knows' principle, meaning that it is hard to predict new trends or anticipate the market success of an innovative design. Maintaining tight-knit professional networks and social contacts with clients and prospective clients helps to minimize this uncertainty. Co-located in localized clusters, design professionals can review one another's products, get to know clients face-to-face, understand their priorities and needs, and keep abreast of the internal politics of their firms. This professional ecology often extends to a broad spectrum of creative industries. Professional interdependencies and the collective nature of creativity means that innovative design depends on cross-fertilization with cultural-products industries (such as art, advertising and branding, photography, publishing, software development, computer and video games, film and video, performing arts and television) and related hybrid and crossover industries (such as event, exhibition, and set design, public relations, tourism and heritage industries). As a result, there is a broader sociality to the industrial atmosphere. As Elizabeth Currid (2007) observed in Manhattan:

> People talked, they compared notes. They changed jobs. And when one engineer or designer meets with another to talk about how a new computer's design will fit with the hardware inside, or whether a particular fabric will work with a designer's spring collection, chances are they exchange a lot of ideas – even ideas not necessarily related to the task at hand, from the names of other pattern makers to what is going on in Milan's fashion industry. That exchange of knowledge ended up translating into new ideas and product innovations.
>
> (Currid 2007: 71)

This aspect of sociality tallies with research in economic sociology, which has emphasized the notion of 'embedded firms' in networks of economic, social and cultural relations (Granovetter 1985, 1991; Crewe 1996). This perspective views economic activity as embedded in networks and institutions that are socially constructed and culturally defined and acknowledges the importance of 'untraded interdependencies' among firms. It offers ways of understanding the various informal arrangements and practices and tacit knowledge that are not fully quantifiable but nonetheless enhance innovation and creativity through networks of actors. In this context, Mark Granovetter's work was seminal in pointing to the importance of 'weak ties' – social and professional acquaintances rather than close friends and colleagues. The relationships most important in professional success and information exchange, it turns out, are not those rooted in familiar, strong trust bonds but rather those involving linkages among people who are not in close connection with one another. Thus, people with large numbers of weak ties find themselves in the best position to diffuse innovation, and the more economic and social opportunities they will have. This was resoundingly confirmed by Currid's (2007) interviews with participants in the cultural economy in New York City. She found that cultural producers rely heavily on their social lives to advance their careers, obtain jobs, and generate value for their goods:

> In the creative economy, weak ties (e.g. knowing that producer or that graphic designer who you met at a party or recognizing mutual acquaintances) carry great weight because within art and culture weak ties are economic opportunity. ... it is the network of people and industries, the editors, magazines, the public relations, and the celebrities that are essential to selling your product – people that are not personally close to you (and not even necessarily in your industry . . .).
>
> (Currid 2007: 78–79)

Job opportunities, along with the embodied cultural capital and tacit knowledge that is so important to practitioners of cultural-products industries, are often heavily reliant on buzz, social contacts and acquaintances. The co-mingling of artists, artisans, designers, photographers, actors, students, educators and writers in cafés, restaurants and clubs and – for some – gallery openings, fashion after-parties, music release events, celebrity birthday parties contributes to a hip, cool 'scene', a blurring of the social worlds of work and lifestyle that is a distinctive dimension of creative-industry clusters in larger cities.

Fabric effects

There is also a 'fabric effect' that contributes to the agglomerative tendencies and dynamics of clusters of design services, and to their overall character. This has to do with the economic and aesthetic attributes of the built environment – the fabric of the city – and it operates in terms of both the commercial and the residential preferences of design professionals and their colleagues in ancillary and crossover industries. Broadly speaking, the preferred setting for many design firms and their professional employees is an inner-city transitional neighbourhood characterized by a combination of heritage buildings, affordable older spaces ripe for conversion or renovation, and a selection of 'real' stores, bars and restaurants with local 'character'.

Typically, such neighbourhoods are also fairly close to downtown theatre and commodified nightlife, as well as to downtown cultural and office districts. They are, as we have seen, also the classic targets for gentrification, and it is often design professionals who are the pioneer gentrifiers. Disused and obsolescent workshops and warehouses are renovated as offices, galleries and bookstore/cafés. Loft spaces are renovated as apartments, while as-yet unimproved housing provides cheap accommodation for students, 'starving artists' and other low-income households. Such settings meet important personal identity needs for many designers as well as for the image and branding of the firms for which they work. By moving into a neo-bohemian scene, designers signal their radical-progressive commitment to themselves and to others, including clients. They have distinct ideas about what living like a designer should entail, and this in turn impacts the affect of the neighbourhood. As noted in Chapter 2, the net result tends toward a neo-bohemia with a distinctive habitus of spaces, places, personal styles, daily practices, comportment and patterns of consumption.

Such districts are subject to rapid change, however (see Case Study 7.1). As more and more better-paid young (and some older) symbolic analysts/advanced services middle classes/upper professionals are attracted to the creative buzz and lower property values of neo-bohemia, less affluent households are displaced to 'new bohemias' in alternate or adjacent transitional districts. Drawn together at high densities in these new bohemias, aspiring artists and young designers, musicians, and unemployed or entry-level cultural-industry workers amount to what Richard Lloyd (2004) describes as a 'cultural proletariat', a reserve army of design-oriented workers with economically self-sacrificing dispositions. Such districts, Lloyd suggests,

> serve as sources of input, both of raw cultural products and exportable talent, into a globalized chain of cultural production and consumption. I argue that they do so in a fashion advantageous to culture industries; the ideological predispositions and material strategies nurtured in the local field are exploitable by corporate behemoths in film, television, and music, as well as by the institutionalized fine arts market.
>
> (Lloyd 2004: 369)

These concentrations of inexpensive talent and expertise are naturally attractive to all sorts of design-related workshops, galleries and offices, thus making them targets for a new round of gentrification (Lloyd 2005).

Case Study 7.1 **Hoxton's double transformation**

In the early 1990s, Hoxton, an inner-city district to the north of London's financial district, began to show signs of a transition from a rather ordinary neighbourhood to an incipient neo-bohemia. Within a decade, Hoxton had achieved international notoriety as an edgy and innovative cultural quarter. But by the early 2000s Hoxton had become associated with a 'bunch of posers' (Pratt 2009: 1056). The frontier of cool had moved on.

Hoxton's emergence as a cultural quarter was partly attributable to the 'fabric effect' of cheap space close to the centre of London. The neoliberal economic legacy of Thatcherism had devastated neighbourhoods like Hoxton:

> The recession of 1989 to 1995 became the engine of change for a city gridlocked during the 1980s by speculative greed. It created a huge variety of vacant spaces that were used for unprecedented levels of artist-led activity. The city became somewhere impecunious artists, curators, designers and DJs could live, work and play.
>
> (Dexter 2001: 73)

This coincided with the emergence of the art movement known as 'Young British Art' (YBA) or 'Brit Art', led by the likes of Damien Hirst, Rachel Whiteread, Sarah Lucas and Tracey Emin. In the early 1990s, Hoxton became the seat of the movement, and it did not take long before the movement became a 'scene':

> Later, bars and clubs opened; for example, a famous and pioneering gay club, the London Apprentice (later '333'), was on the corner of Hoxton Square. The YBAs socialized in the Bricklayers Arms on Charlotte Street. The Lux cinema, an arts cinema and the home of the London Film and Video Workshop, took space in Hoxton Square and it became a social and artistic hub.
>
> (Pratt 2009: 1047)

Figure 7.2 Renovated workshops in Hoxton. (Photo: author)

Soon, the buzz generated around the sociality of the neighbourhood began to attract an avant-garde of graphic artists, interior designers, photographers, architects and media workers who colonized the old furniture workshops of the area (Figure 7.2). Andy Pratt (2009: 1047) notes: 'Hoxton was crowned one of the "coolest places on the planet" by *Time* magazine (1996) and it was linked to the notion of "Cool Britannia" a theme exploited by the incoming Labour administration in 1997.' Hoxton had developed a distinctive habitus that pioneer gentrifiers found attractive, so much so that property prices began to rise, prompting many of the artists and neo-bohemians to leave for cheaper space further east, around Brick Lane and, further still, Dalston.

Key readings

Andy Pratt (2009) 'Urban Regeneration: From the Arts "Feel Good" Factor to the Cultural Economy: A Case Study of Hoxton, London', *Urban Studies*, 46.5–6: 1041–1061.

Aidan While (2003) 'Locating Art Worlds: London and the Making of Young British Art', *Area*, 35.3: 251–263.

Communities of practice

A great deal of design work is undertaken project-by-project, with different sets of actors, firms and institutions coming together temporarily for the completion of a particular task. Over time, networks and relationships evolve into 'project ecologies' with intricate interdependencies between temporary collaborators and permanent ties, and diverse professional imperatives with some complementarities and some conflicts (Grabher 2004; but see also Sunley *et al.* 2008). At the fringe of these project ecologies, cultural industries exhibit a relatively high degree of self-employment, casualization, 'self-exploitation' and chronic job insecurity. In large cities, the project ecologies of designers and associated ancillary and crossover professionals operate as 'communities of practice' – networks of individuals whose lives are bound together through multiplex day-to-day relationships, based on the same sets of expertise, a common set of technological knowledge, and similar experience with a particular set of problem-solving techniques (Wenger 1998):

> Communities of practice thus lead to the generation of distinct routines, conventions and other institutional arrangements. . . . Clusters can become

important catalysts for the formation of such communities. In this case, they
develop into local frames to understand the meaning and significance of local
buzz which in turn serves to stimulate the generation of local buzz and its rapid
diffusion.

(Bathelt *et al.* 2004: 39)

In broader context, the emergence of communities of practice and the consequent
increase in localized sociality and buzz can be seen as further reinforcement of
cumulative causation.

Central to the operation of communities of practice are the formal and informal
institutions that support design cultures. Just as Becker's 'art worlds' depend on
the presence of patrons, dealers, critics, gallery owners and collectors, auction
houses, critics and art schools, so communities of design practice depend on cultural intermediaries, gatekeepers and tastemakers (columnists, editors, magazine
writers and bloggers for sites like Designboom, Greenopia, Bldgblog and The
Sartorialist) and on complementary institutions such as design schools, technical
colleges, specialized book stores, exhibition spaces, public agencies, foundations,
trusts and community-based organizations committed to fostering design activities
and craft-based skills. Design schools provide not only formal training but also an
interface between industry (through internships), the professional avant-garde
(through lectures and visiting teaching assignments) and the media (through
magazine articles). It is also in design schools that students develop the social
networks and key contacts that will underpin the critical sociality of 'weak ties' in
their own particular branch of the industry. Design-related events such as gallery
openings, catwalk shows and award ceremonies also play an important role, both
in terms of highlighting work and providing informal networking opportunities for
career advancement and access to gatekeepers. As Thomas Hutton (2004: 93)
observes, the density and 'thickness' of all these institutional networks represent
critical underpinnings not only of specific communities of practice but also of the
new economy as a whole.

'Place in product'

Distinctive design cultures can emerge as a result of the particular mix of local
cultural and design-related activities and the local taste systems of producers and
consumers, along with the general influences of specific national and regional
cultures. The relationship between locality and design is recursively reinforced
through what Harvey Molotch (2002) calls the 'valorisation of milieu'. The result
is 'place in product', the way that locality comes through in product design,
architecture, and urban design. As Molotch puts it:

Products depend on just how the expressive, the material, and the organizational elements connect in a given place at a given time. People breathe in their industrial atmosphere through the stories, jokes, manners, architecture, street styles, sounds, odors and modes of maintenance that surround, as well as what they remember and what they anticipate. They enroll in particular projects with special attention and even gusto because they have the local means to 'get it', to fit in, to function as a more or less reliable member. . . . Because local character, tradition and difference arise from face-to-face interaction and in situ cultural absorption, place differences endure in their productive consequences. . . . It is the idiosyncratic ways that places work that color what a thing can be. The various spheres come together not because they are intrinsically related, but because the nature of the place causes them to be related – and in a particular way, for a given moment. Place gets into goods by the way its elements manage to combine, and the stuff shows it.

(Molotch 2002: 685–686)

With food products and manufactured goods, place in product can be seen to operate at the regional or urban scale, as in California wine, Parma ham and Sheffield steel. With many aspects of design, however, place in product is articulated at much broader, national or continental scales. Thus, for example, we talk of 'Euro design', 'Scandinavian design' or 'Italian design'. 'Euro design' has become shorthand for contemporary Modernism that echoes the clean lines and functionality of Bauhaus – but with a distinctly up-market appeal that contrasts with the socialism of the original movement. Scandinavian design derives its place-in-product qualities from the mid-twentieth century architecture and product design of Alvar Aalto, Arne Jacobsen, Eero Saarinen and Timo Sarpaneva. It resonates with Scandinavian social democracy and understatement; and it has, of course, been commodified and globalized by Ikea.

Italian design arguably has the strongest overall place-in-product associations. This is partly due to the reputations of leading figures in the Milanese design cluster (including Ferdinando Innocenti, Renzo Piano, Sergio Pininfarina, Giò Ponti, Aldo Rossi and Ettore Sottsass). Their success, in turn, rests in part on the success of innovative, high-design Milan-area manufacturers such as Alessi, Alfa Romeo, B&B Italia, Cassina, Kartell and Lambretta, and fashion houses such as Armani, Dolce & Gabbana, Gucci, Prada and Versace. The tradition of specialized, innovative and high-quality craft industries in Lombardy has been important to all of these firms, while the general importance of *la bella figura* in north Italian culture has ensured a strong base of consumer support and media interest.

It has been claimed that the British approach to design tends to be slightly anarchic and irreverent (Sunley *et al.* 2008; O'Byrne 2009). In addition, ethnic influences – partly from Britain's ties to former colonies – have been associated recurringly with British fashion design (Breward and Gilbert 2006). The place-in-product dimension here centres on the openness, radical disposition and cultural diversity of the London 'scene'. But for London, as for the other largest world cities, the distinctiveness of place in product tends to be trumped by the sheer size and variety of its cultural clusters. The city itself, like New York and Paris, has become established as a global tastemaker, the 'valorisation of milieu' extending to all sorts of products and activities through branding that simply invokes the city's name. At the same time, it is clear that the communities of practice of these world cities draw on professional and social networks that extend well beyond individual intra-urban clusters. Global 'pipelines' have become increasingly important as design-based goods and design services are increasingly articulated in *filières* (chains of creation of value) that are international in scope. Here is yet another component of cumulative causation: local clusters and global pipelines are mutually reinforcing:

> The more firms of a cluster engage in the buildup of translocal pipelines the more information and news about markets and technologies are 'pumped' into internal networks and the more dynamic the buzz from which local actors benefit.
>
> (Bathelt *et al.* 2004: 41)

We turn now, therefore, to the globalization of design services and the role and dynamics of design services in the political economy of world cities.

The globalization of design services

As we have seen, professional practice in design has long had an international component and a cosmopolitan outlook. The international reach of architecture as a profession was established well before the Second World War by the commissions of leading practitioners such as Albert Kahn and Le Corbusier (Scully 1988; Frampton 1992). It was consolidated by Philip Johnson and Henry Russell Hitchcock's promotion of the idea of an 'International Style'; by the international migration of Bauhaus members, who applied their ideals to product design as well as architecture; by the publication of the Athens Charter developed by CIAM (the Congrès Internationaux d'Architecture Moderne); and by the colonial practices of British, French and Italian architects. After the Second World War it was further propagated by the commissions of successive generations of leading practitioners; and, more prosaically, by some large US architecture and engineering firms, such as CRS, whose commissions derived from the US government's foreign aid projects (many of them focused on infrastructure projects and 'tied' to the participation of US firms) and from the neocolonial investments of US corporations.

But until relatively recently most design practices have been organized around a local, regional or national framework. Globalization has changed all that. Enabled by digital and telecommunications technologies, by advanced international business services, and by the emergence of clients with transnational operations and a cosmopolitan sensibility, the portfolio of many firms has an international component and the scope of operations of many of the largest firms is now truly global, with multiple international offices covering several continents. The World Trade Organization has attempted to codify international trade in design services through the General Agreement on Trade in Services, and imports and exports of design services have increased dramatically (Tombesi 2003).

In many ways, this globalization of design services has been an inevitable consequence of the earlier globalization of manufacturing and of advanced business services. The first wave of economic globalization, in the 1970s, was led by manufacturing giants like General Motors and General Electric, whose global reach had the triple objectives of reducing labour costs, outflanking national labour unions, and increasing overseas market penetration (Knox *et al.* 2008). In the 1980s, the digital technologies of the emerging 'new economy' enabled transnational corporations (TNCs) to draw on new knowledge and to access and process information on finance, commodities, markets and consumer preferences in different regions.

With this development, the leading firms in advanced business services – accountancy, advertising, banking and law – were able to establish a global clientele. The leading business service firms themselves became 'global brands' whose integrity required protection. This could not be guaranteed through arrangements with local firms, so they expanded their office networks in order to provide worldwide servicing capabilities. The literature on the globalization of advanced business services shows that although different sectors such as accounting, advertising, banking and law have different clienteles and different market imperatives, the same major cities – 'world cities' – have become key nodes for the networks of business that drive and shape globalization (Knox and Taylor 1995; Short and Kim 1999; Sassen 2001, 2002; Taylor 2004).

Design firms have been slower to go global, but the sustained economic boom of the 1990s saw many larger firms extend their operations through office networks that are international in scope. The economic logic of globalization had induced an international outlook that is manifest in two practice developments. First, many US- and European-based firms had begun to take advantage of international outsourcing, drawing on pools of skilled but inexpensive labour in South Asia and the Pacific Rim. Here, the division between routine production (drawing and documentation) and symbolic analysis (design and management) is critical

(Tombesi 2003; Tombesi *et al.* 2003). Many European and North American architecture firms, for example, are turning to international outsourcing for working drawings and three-dimensional visualizations – partly because of the reduced costs, and partly because of the increased speed of production when routine tasks are outsourced to overseas firms for overnight completion.

Second, with an increasingly international clientele, firms developed global office networks to serve an increasingly complex market. As with business services, the global strategies of many major design firms have been very client-led. For instance, the San Francisco-based architecture and environment consulting company *EDAW* has been explicit that its location strategy is a response to the global nature of their clients' business. Similarly, the Japanese company *Kajima Design* recorded a significant increase in international operations in the 1980s and subsequently reorganized by setting up regional subsidiary companies in response to a growing customer base across the world; while *CH2M Hill* established 200 offices around the world because 'personal relationships are important for results' (Knox and Taylor 2005).

Meanwhile, as Paul Kennedy (2005) has observed, the emergence of TNCs and global business service firms has generated a lot of big contracts for the design of signature corporate buildings:

> These must simultaneously act as signifiers of corporate uniqueness, power and world stature while serving as administrative centres for TNC national, macro-regional or global operations. . . . The need for prestigious building-design projects comes not only from TNCs but also from the national and international 'public sector' – perhaps the German Federal government, Melbourne city council or the World Bank – involving projects such as museums, art galleries, railway stations, airports or governmental buildings. To be taken seriously for such landmark building contracts, companies must demonstrate a worldwide reputation for excellence.
>
> (Kennedy 2005: 183)

This of course, places a premium on the services of 'starchitect' practitioners (McNeill 2009).

World cities and their ecologies of design

The growth of world cities has meant that there has been plenty of other important architectural work to go around, and this has fuelled the growth of design firms with a global brand and a truly global reach, with offices on several continents. Such firms are localized in major world cities, close to their transnational clientele. In detail, the greatest concentration of global practices, by far, is in London. Other premier architectural practice cities include Hong Kong, Los Angeles, Melbourne, New York, San Francisco, Singapore, Tokyo and Washington, DC (Knox and Taylor 2005). There are also global-practice offices in several cities of the Middle East and Pacific Asia (Figure 7.3). This basic geography of globalized architectural practice confirms that this particular dimension of globalization is very much market led, though the regional markets are very different: US cities represent a long-term tradition of large-scale building renewal and development; Pacific Asia has been the boom region of contemporary globalization with development focused upon its major cities; Australia represents a smaller version of the US cities process; and the Middle East is a region of concentrated wealth creating political clients who are trying to boost their local cities. London, meanwhile, is clearly *the* place to be for global architectural practice, and its pre-eminence seems to have cast a 'shadow effect' on other Western European cities. Paris, in particular, is dramatically underrepresented in comparison to its share of global business services.

Fashion design is also localized in a few world cities, and here the spatial pattern reflects a different story, with Paris having a long-established pre-eminence in couture that has coloured the affect and shaped the imagined geography and media representation of the entire city (Rocamora 2009: 85). With the emergence of prêt-à-porter (ready-to-wear) and the expansion of global consumer markets, fashion has become an important feature of global competition among cities, part of broader strategies of metropolitan boosterism and a catalytic element of intra-urban clusters of design services (Breward and Gilbert 2006). The list of cities that hosted major catwalk shows in 2009 includes established world cities such as Berlin, London, Los Angeles, Milan, New York, Paris, Seoul and Sydney as well as aspiring cities such as Copenhagen, Dublin and Stockholm (Figure 7.4). Fashion weeks have become critical to reinforcing claims to cosmopolitan 'world city' status. They bring together a diverse range of interests and specializations (Weller 2008) – event organizers, fashion designers, fashion retailers and wholesalers, clothing manufacturers, textiles makers and designers, the fashion media, as well as cosmetics, personal care and hospitality services, and other specialized fashion intermediaries – and add momentum to the creation and capture of value through circuits of capital that pivot around luxury consumables, media products and real estate, as well as the garment industry and design services (Figure 7.5).

The magnetism of world cities to design services hinges on such circuits of capital. As talented designers are drawn from elsewhere, the buzz of clusters, the potency of branding and the momentum of cumulative causation are all reinforced. The history, character, composition and spatial organization of design services vary from one city to another, however, as illustrated by three of the most prominent world cities in terms of design: Paris, New York and Milan.

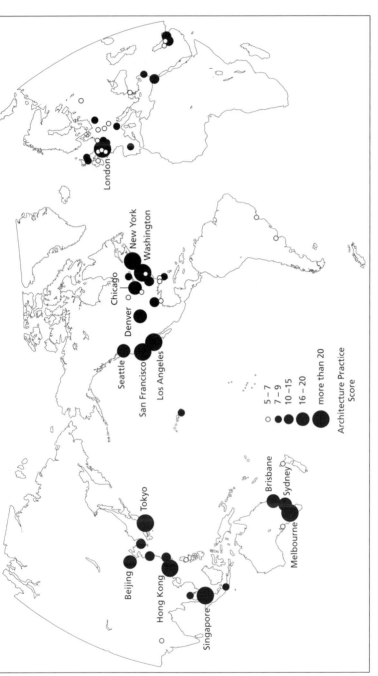

Figure 7.3 Major concentrations of architecture firms with global practices. The 'architecture practice score' for each city is based on the presence of offices of firms with a network of offices on at least three continents. A score of 5 is allocated for the firm's headquarters; cities with ordinary or typical offices are allocated 2; minor offices are then identified and score 1; major offices score 3 or 4. (Source: After P.L. Knox and P.J. Taylor (2005) 'Toward a Geography of the Globalization of Architecture Office Networks', *Journal of Architectural Education*, 58:3: 23–32)

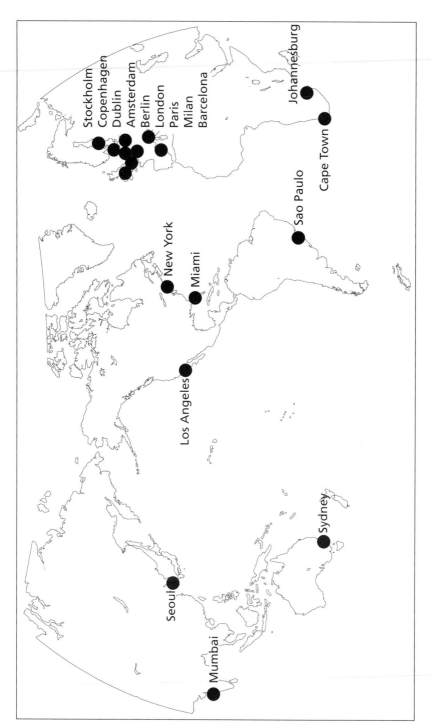

Figure 7.4 Cities with catwalk shows of major fashion collections in 2009, as recognized by *Vogue* magazine.

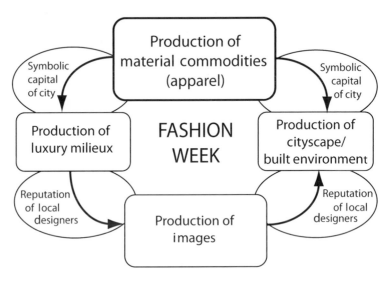

Figure 7.5 Flows of value surrounding fashion week events. (Source: After S. Weller (2008) 'Beyond "Global Production Networks": Australian Fashion Week's Trans-Sectoral Synergies', *Growth and Change*, 39.1: 112)

Paris

Paris has long enjoyed a reputation as a global tastemaker, a city of the arts that is often portrayed as a world of luxury goods, glamorous haunts and fashionable lifestyles. Although not a preferred location for architecture practices with a global reach, it does have, like other major world cities, a large and diverse array of design services and related cultural-products industries. Foremost among them is the fashion industry, whose strength can be traced to the emergence of Paris as the 'capital of modernity' in the nineteenth century.

As we have seen (Case Study 1.1), the city itself was remade as an object of desire and consumption, 'a city enjoying itself in the midst of a profusion of goods and cafés' (Rocamora 2009). The *passages*, department stores – *Printemps, Le Bon Marché* – and the gardens of the city – Parc Monceau, Jardin des Tuileries, Luxembourg Gardens – became settings for the public display of high fashion, a key aspect of the new consumerism. Charles Frederick Worth, an English designer, moved to Paris in 1848 and effectively created the first designer brand (the House of Worth), reinventing the role of the couturier as arbiter of style and taste. A rash of books on Paris contributed to framing the city 'as an object of

desire, a site of prestige and a place of sartorial elegance and fashionable display'
(Rocamora 2009: xv).

By the early twentieth century, thanks largely to the availability of fashion imagery
and the introduction of factory-based production, there had developed a democ-
ratized international fashion system, with Parisian style as the principal point of
reference. The establishment of *La Chambre Syndicale de la Couture Parisienne*
in 1910 represented the beginning of a formal institutionalization of the industry,
coordinating organized fashion shows and seasonal collections (though the first
catwalk shows with live models and music were criticized in *L'Illustration* as part
of a 'hideous crisis of bad taste': Gilbert 2006: 22).

After the Second World War, the emergence in Western societies of a 'romantic
capitalism', driven by the 'self-illusory hedonism' of dreams, fantasies and competitive consumption (see Chapter 1) coincided with the commodification of elite
designers' reputations and the direct licensing of prêt-à-porter fashions labelled
with designer names. Other cities also began to organize fashion weeks and to
promote local design, beginning with Florence in 1952 and London in 1958. Paris
became increasingly reliant on the mystique of haute couture and the potency of
its symbolic capital. The attraction was sufficient to draw in the likes of Elsa
Schiaparelli and Christòbal Balenciaga among others, and the assembled design
talent was able to draw on the skills and flexible specializations of scores of small
workshops that remained in the le Sentier district of the 2nd arrondissement, close
to the central business district in the heart of the city.

In the 1980s, as urban entrepreneurialism and city branding became more
competitive, the Parisian fashion industry was vigorously promoted by the French
government. Designers were given honorary decorations, and the institutional
infrastructure of the design industry was promoted. The Institut Français de la
Mode, a high-profile postgraduate management school specializing in fashion,
was opened in 1986, supplementing the well-established design schools at Studio
Berçot and l'Ecole de la Chambre Syndicale de la Couture. More recently, the
Institut Français de la Mode has moved to the Cité de la Mode et du Design in
the 13th arrondissement, a signature structure that also contains design ateliers,
stores, restaurants, cafés and event spaces. The fashion press has also been central
to maintaining Paris's brand as *capital de la mode*, tightening the relation between
fashion and urban culture, between semiotic goods and designer labels, and
between the city's name and fashionability.

In Paris, as in other cities, there is a marked spatial separation between the
production spaces and the consumption spaces of design (Figure 7.6). In the case
of fashion – especially women's wear – production has remained in the Quartier

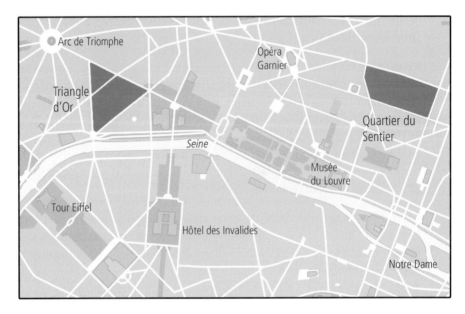

Figure 7.6 Fashion districts in central Paris.

du Sentier, where design ateliers are crowded in with couturiers' workshops as well as firms producing affordable prêt-à-porter. The 'front region' of fashion design, its principal consumption space, is the 'Triangle d'Or', stretching between the Avenue des Champs-Elysées, the Avenue Montaigne and Avenue George V in the 8th arrondissement. Here are the offices and showrooms of the most celebrated Parisian couturiers – Chanel, Dior, Nina Ricci, Ungaro – the cornerstones of a 'semiotic district' (or, as Agnès Rocamora (2009: 50) calls it, a *griffe spaciale* – a 'griffe', in the language of fashion, being the designer label affixed to a branded product).

New York

In broad terms, the emergence of New York as a centre of design services in the early twentieth century was related to the realignment of global capitalism and its role as the corporate and commercial hub of a rising economic superpower. Concentrations of corporate headquarters, banking, publishing and advertising generated an affluent market for many aspects of design, especially architecture, graphic design, interior design and fashion. As an emerging world city, New York also attracted other cultural-products industries, artists and musicians. Today, as one of the most dominant world cities, New York also stands at the forefront of 'designer cities'. Indeed, Elizabeth Currid (2006) goes as far as suggesting that

New York City's real competitive advantage and unique position as a world city rest today on its setting as a great centre of creativity in the arts, design, media and entertainment (Currid 2006).

The 'fabric effect' of Manhattan's geography is important, providing a range of neighbourhoods with affordable older spaces ripe for conversion or renovation, a broad spectrum of office space, a variety of gentrifying neighbourhoods, neo-bohemias, semiotic districts and vibrant nightlife. Not surprisingly, the institutional framework supporting design and cultural-products industries in New York is well developed, including, notably, the Museum of Modern Art, the Solomon R. Guggenheim Museum, the Cooper-Hewitt Design Museum, the Pratt Institute, Parsons School of Design, the New York School of Interior Design, the Fashion Institute of Technology, the Council of Fashion Designers of America, Columbia University, the showrooms of the New York Design Center, and gatekeepers and tastemakers at the likes of *Architectural Record*, *Harper's Bazaar*, *Interior Design*, *Metropolis* and *Vogue*.

Within this broad socio-cultural ecology, there are several design-oriented clusters (Figure 7.7). Fashion design (see Case Study 7.2) and interior design are localized in the Garment District (recently re-branded as the Fashion Center), with nearby Bryant Park the setting for the catwalk shows of New York Fashion Week. The Lower East Side is a secondary production space for the fashion industry and also the setting for myriad small design and design-related firms. Larger architectural practices are clustered in the midtown office district, with graphic design drawn toward the locus of advertising agencies on Madison Avenue. Software design and new media are clustered nearby, around the intersections of Fifth and Sixth Avenues, and 18th and 21st Streets: 'Silicon Alley', a label promoted by the New York New Media Association (Pratt 2000). Smaller architectural practices, along with smaller interior design, graphic design and product design studios, tend to be localized in SoHo, Chelsea and the Meatpacking District, as well as the Lower East Side, the districts that also facilitate the buzz and sociality in the galleries, cafés, bars, clubs and restaurants that Currid found to be so important in New York.

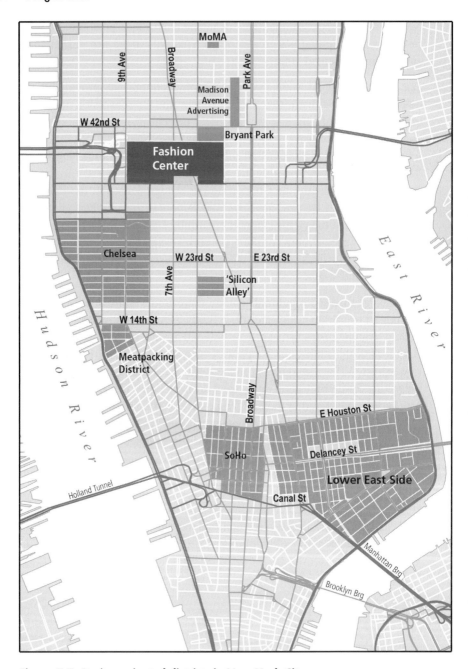

Figure 7.7 Design-oriented districts in New York City.

Case Study 7.2 **New York's Fashion Center**

The Fashion Center in New York is the branded neighbourhood traditionally known as the Garment District. It is bordered to the north by 40th Street, to the south by 34th Street, to the east by Fifth Avenue, and to the west by Ninth Avenue. In contrast to the haute couture origins of the fashion industry in Paris, New York's fashion industry has its origins in ready-to-wear. It first emerged in the mid-1800s to meet a surge in consumer demand as the United States was experiencing its first signs of industrialization and urbanization. As geographer Norma Rantisi (2002b) observes:

> Its rise and expansion coincided with the invention of the sewing machine in 1846, which allowed for volume production, and a major wave of immigration from Southern and Eastern Europe in the 1880s, consisting primarily of Jews and Italians, which provided a ready supply of skilled labor.
>
> (Rantisi 2002b: 592)

New York's women's wear industry was subsequently boosted by the closure of Paris as a result of Nazi occupation in 1940. Then, in the postwar economic boom, there was a surge in domestic demand for ready-to-wear, including the huge new markets for leisure wear and designer sportswear. Brand-name designers like Bill Blass, Donna Karan, Calvin Klein and Ralph Lauren were closely associated with New York, but the Garment District itself remained rather unglamorous, a tight-knit agglomeration of fashion-related enterprises in ageing workshops, offices and showrooms (Figure 7.8).

As the more upmarket fashion retailers began to move out to SoHo, and as some manufacturers moved out of New York altogether in search of cheaper labour, the Garment District was left to evolve into more of a design centre than it had ever been (Merkel 2000). Rebranded as the Fashion Center (Figure 7.9), it still contains many small manufacturers and specialized retailers. According to Norma Rantisi (2002a), the four-by-six-block area accounted for $17 billion in apparel manufacturing and wholesale output in the early 2000s. About 6,000 firms were packed into the area, of which

> roughly 4,000 are fashion-related businesses, which – in addition to apparel manufacturers and contractors – include retailers, textile suppliers, resident buying offices, forecasting services, trade publications, and fashion design schools, as well as a variety of legal, financial, and supply services.
>
> (Rantisi 2002a: 448)

**Figure 7.8
Factory buildings in
the Garment District.
(Photo: author)**

Rantisi argues that it has been the variety as well as the economic coherence of the District that has produced a design innovation system that has sustained its ability to adapt to shifting competitive pressures.

There are signs, however, that the manufacturing base of the Fashion Center is under threat. Since 1987, the city has protected the Garment District through special zoning that restricts landlords from converting factory space to offices and budget hotels, which command higher rents. To appease landlords, city officials wanted to rezone the Garment District in 2009 and move the remaining clothing manufacturers to Brooklyn or Queens. Industry leaders, along with the Fashion Center Business Improvement District and the Council of Fashion Designers of America, have argued that without production in the Garment District, there would be no reason for designers and suppliers to cluster there, where they can walk to sample rooms, visit pattern makers and drop in on factories to oversee production quality. The likely compromise will be to identify and protect a minimal core of factory space while eliminating the overall zoning restrictions in the rest of the District.

Figure 7.9 Donna Karan plaque on the Fashion Center's walk of fame. (Photo: author)

Key readings

Elizabeth Currid (2007) *The Warhol Economy: How Fashion Art and Music Drive New York City*. Princeton, NJ: Princeton University Press.

Norma Rantisi (2002) 'The Competitive Foundations of Localized Learning and Innovation: The Case of Women's Garment Production in New York City', *Economic Geography*, 78.4: 441–462.

Norma Rantisi (2006) 'How New York Stole Modern Fashion', in Breward, C. and Gilbert, D. (eds) *Fashion's World Cities*. Oxford: Berg.

Milan

It does not take long for visitors arriving in Milan by rail or air to pick up on the idea of Milan as a designer city: they are confronted almost immediately with giant photographs of airbrushed fashion-forward teenagers (Figure 7.10), reminders of Milan's most recent economic miracle. After the Second World War, Milan boomed as Italy's leading industrial centre. Hundreds of thousands of jobs were generated by the growth of steel works, chemical refineries, car factories, engineering works and rubber factories. But the heavy industrial boom was brief, and Milan soon began to feel the effects of deindustrialization. The decline took place before large-scale industrialization had completely obliterated the artisan workforce and the workshops of the specialized, small textile and furniture businesses of the city and small nearby towns such as Biela (wool), Carpi (knitwear), Castelgoffredo (hosiery), Como (silk) and Vigevano (footwear). These, as Andrea Branzi (1994: 239) notes, were 'the historic cradle of Italian design'. By the late 1950s, Milan had acquired a reputation as a global capital of industrial design, and the flexible specializations of the textile and furniture workshops were critical:

> The reasons behind the rapidity and duration of this remarkable transformation lie in the particular combination of intellectual and professional milieu (especially in the realm of architecture, industrial innovation and concentration and market capabilities centred around Milan and Lombardy). Architects, designers, and architect-designers were able to work directly within the rich

Figure 7.10 Dolce & Gabbana posters at Milano Centrale train station. (Photo: author)

fabric of small industry, artisans, and workshops that characterized (and still characterizes) the vast productive region to the north of Milan. In addition, the showcase possibilities provided by the international Triennale exhibitions of the 1940s and 1950s brought Italian design (and Milanese design) to world attention.

(Foot 2001: 110)

Guy Julier (2000) points to another factor: a renewed ideological focus on the home 'as a site for the personal, moral, and social reconstruction following the failure of fascism' in Italy. Yet professional design culture was also geared toward international peer recognition. The net result, writes Penny Sparke (1990: 194), was that modern Italian design became oriented to 'a predetermined "culture of the home", associated with a particular middle-class, consumerist lifestyle . . . turning all Italian objects . . . into highly desirable, élitist artifacts'.

This was the era of Alfa Romeo cars, Bianchi bicycles, Gaggia espresso machines, Lambretta scooters and Valextra briefcases. Exceptional designers were matched by an exceptional group of entrepreneurs like Carlo Alessi, Piero Businelli (of B&B Italia), Cesare Cassina, Giulio and Anna Castelli (of Kartell) and Gino Colombini and Joe Colombo (of Artemide). Milan's affluent industrialists, meanwhile, were determined to leave their mark on the city. The Pirelli Tower, designed by Giò Ponti and opened in 1960, was Europe's first convincing skyscraper and quickly became a symbol of the new Milan, a signal of confidence and faith in the future of the city.

Milan also had a strong foundation for the institutional infrastructure of design. The prestigious design school, Istituto Marangoni, was founded in 1935. The triennial exhibitions of art, architecture, design and planning at the Triennale that began in the 1920s were planned to coincide with conferences and experimental architectural projects and brought international attention to Italian design. The editors of *Domus* were responsible for inventing 'Italian design' for an international audience while *Casabella* laid the basis for a distinctive approach to urban design. These influential magazines were joined in the 1960s and 1970s by Milan-based *Abitare*, *Casa Vogue*, *Modo* and *Ottagono*, while the city's flagship department store – *La Rinascente* – sponsored the *Compasso d'Oro*, Italy's top design prize. The store also made a point of hiring fresh design talent, including Giorgio Armani, who worked in the store's fashion department 1954–1960, first organizing the shop-window collections and then as a fashion buyer. The annual Fiera di Milano (formerly the Fiera Campionaria) and Salone Internazionale del Mobile became internationally renowned as a showcase for architects, interior designers, and the designers and manufacturers of furniture and lighting (Romanelli 2008).

Milan's status as a designer city was cemented in the 1970s when it hosted the first fashion shows, breaking with the suffocating traditionalism of Florence. The meteoric rise of Milanese designers – Giorgio Armani, Stefano Dolce, Gianfranco Ferré, Elio Fiorucci, Domenico Gabbana, Miuccia Prada and Gianni Versace – and their prêt-à-porter brands meant that the city became a magnet for photographers, models, magazine editors, critics, buyers, manufacturers, commercial traders and journalists (Reinach 2006). A key development here was the alliance in the 1980s between the fashion industry and the city's socialist administration (*Partito Socialista Italiano* – PSI). Political scientist John Foot (2001) notes:

> Many fashion designers worked directly for PSI politicians and in return received the go-ahead for a series of economic and urban projects, and permission to hold fashion shows in a wide range of urban institutions, from La Scala to the Triennale to the Stock Exchange to the Racing Track to the Central Station.
>
> (Foot 2001: 129)

By the 1990s:

> Milan and its surrounding territory constituted a design system. This system was made up of private and public institutions, industries, magazines, designers and studios and a series of services linked to production and advertising of design goods and ideas. The centre of this system was the annual Salone del Mobile, but design events took place in the city throughout the year, attracting international interest.
>
> (Foot 2001: 124)

Not least among these events are now the Women's and Men's Fashion Weeks, which dominate Milanese commercial life.

Milan's success as a global centre of design is reflected in its infrastructure and built environment. Milan's private university, Libera Università di Lingue e Comunicazione (IULM) has established a faculty of design, and its prestigious private business school, Bocconi, offers specialized, fashion-oriented Master of Business Administration degrees. The Triennale building now also operates as a design museum, and a new branch of the Triennale has been established in the deindustrialized district of Bovisa, where the Politecnico di Milano has built a branch campus on a brownfield site that features its Design Faculty. An enormous new Fiera complex, designed by Massimiliano Fuksas, has been built on the site of a former oil refinery at Rho-Pero on the edge of the city. Its curving glass roof sweeps over a central congress area (designed to cope with half a million visitors) and then runs in a ribbon down a central 'street' that is over a kilometre long. This has freed up the Fiera's valuable former site, now being redeveloped as 'Citylife', a mixed-use development designed by a combination of architects that includes Daniel Libeskind, Zaha Hadid and Arata Isozaki.

The signature affect of contemporary Milanese fashion and design (apart from the borderline kitsch of Versace) is Modernist-flavoured elegance and luxury: ideally suited to the 'dream economy' of the 1980s and 1990s. Versace, of course, fills nicely the market niche of the dream economy that is upscale bling. This affect is very much in evidence in fashion's 'front region', the 'Quadilatero d'Oro', a well-to-do area on the eastern fringe of the central business district formerly dominated by antique dealers. Clustered around the four streets of Via Montenapoleone, Via della Spiga, Via Sant'Andrea and Corso Venezia, the Block of Gold contains several hundred upscale fashion outlets, many of them unwelcoming and exclusive (Figure 7.11).

In contrast, the production-oriented district around Porta Ticinese to the south of the city centre has a distinctly neo-bohemian affect. Since the mid-1980s the centre of gravity of design services has migrated slowly eastwards towards Porta Genova, where it has become branded as a self-consciously design-services district under the label Zona Tortona (Figure 7.12). Formerly a warehouse and small-scale furniture fabrication district, Zona Tortona's transformation began in the mid-1980s, when Flavio Lucchini and Fabrizio Ferri set up 'Superstudio' using a renovated former bicycle factory as a fashion photography studio. Other photographic studios soon appeared, followed by fashion offices and showrooms of global brands such as Armani, Zegna and Tod's, and architects' and industrial designers' offices, sharing the same area with the local furniture and furnishings

Figure 7.11 Shoppers in Milan's Quadilatero d'Oro. (Photo: author)

Figure 7.12 The Sportswear Company (SPW Company, including the brands CP Company and Stone Island) buildings in the Zona Tortona, occupying the former factory, trade union building and cafeteria of an engineering works. (Photo: author)

workshops. With the galleries, bookshops, trendy restaurants, bars and cafés that have also appeared (or been revamped), Zona Tortona now represents one of Europe's most vibrant communities of practice, with a sociality that encompasses models, photographers, art directors, film directors, stylists, crafts workers, industry workers and small shopkeepers as well as designers of all kinds.

Globalization and designer cities

Paris, New York and Milan are just three examples of 'designer cities'. London and Tokyo, as major world cities, can claim similar status, though with distinctive histories, composition and spatial organization of design services. Other cities, too, can point to a significant 'designer' dimension to their economic profile. Design companies in Berlin, for example, generate annual sales of over €1.5 billion, while almost 12,000 Berliners work in fashion, product and furniture design, architecture, photography and the visual arts and about 5,000 students are enrolled in fashion, product, graphic and communication design, photography, or visual arts programmes.

As noted earlier in this chapter, the professional and social networks in cities like Berlin, London, Milan, New York, Paris and Tokyo extend well beyond individual cities. Global 'pipelines' have become increasingly important as design-based goods and design services are articulated in *filières* that are international in scope. Although perhaps not on the same scale as in major world cities, there are agglomerations of design services in many cities, drawn together through the internationalization of design media and education and by the increasingly globalized aesthetics of design and linked together by project-based work involving international clients and partners. At the same time, the potency of design in strategies of urban regeneration, the increasing role of design services in adding value (and profit) within the 'dream economy', the increasing importance of city branding, and the seductive idea of promoting urban economic growth through attracting a 'creative class' means that more and more cities are actively promoting design in one way or another and competing for the symbolic capital afforded by design. Among the cities hosting a 'Design Festival' in 2008–2009 were Aberdeen, Bangkok, Bristol, Cardiff, Cologne, Hamburg, Liverpool, Melbourne, Pune, Sydney and Toronto. Others, including Atlanta, Belgrade, Colombo (Sri Lanka), Copenhagen, Eindhoven, Helsinki, Istanbul, Kuala Lumpur, Manila, Phoenix, San Francisco, Tokyo and Vienna, hosted a 'Design Week'. Auckland, New Zealand, has established a Fashion Week that explicitly attempts to leverage the symbolic cachet of London, New York and Tokyo (Larner *et al.* 2007).

Others have focused on the infrastructure and built environment. Johannesburg has earmarked R35 million (£26 million) to establish a branded Fashion District;

Antwerp has opened Mode Natie, a multifunctional building that houses a Fashion Academy, a Fashion Museum and the Flanders Fashion Institute (Martínez 2007); and Seoul – the 2010 'World Design Capital' of the International Council of Societies of Industrial Design – is building the Dongdaemun Design Plaza, designed by Zaha Hadid, which will incorporate a design museum and design library as well as designer-oriented office and retail space. All this speaks to the central role of design in supporting and sustaining the political economy of urbanized capitalism: facilitating the circulation and accumulation of capital, helping to stimulate consumption through product differentiation, and contributing to social reproduction, legitimation and identity. More than ever before, design is a fundamental dimension of urbanization.

Summary

Big-city environments are forcing houses of cultural innovation and important arenas for the creation of taste. Most large cities contain specialized districts that act as 'creative fields', distinctive settings that are rich with innovative energy, dense interpersonal contacts and informal information exchange. The globalization of design services has intensified these creative fields, especially within world cities. The globalization of design services has been a consequence of the earlier globalization of manufacturing and of advanced business services. Design firms were relatively slower to go global, but the sustained economic boom of the 1990s saw many larger firms extend their operations through office networks that are international in scope. The growth of world cities has meant that there has been plenty of work to go around, and this has fuelled the growth of design firms with a global brand and a truly global reach, with offices on several continents. Such firms are localized in major world cities, close to their transnational clientele.

The interdependence of design with other creative activities and with the broader political economy means that the design professions tend to exhibit high degrees of geographic concentration within cities, especially in world cities. These agglomerative tendencies and dynamics are reinforced by some important aspects of sociality. Job opportunities, along with the embodied cultural capital and tacit knowledge that is so important to practitioners of cultural-products industries, are often heavily reliant on buzz, social contacts and acquaintances. There is also a 'fabric effect' that contributes to the agglomerative tendencies and dynamics of clusters of design services, and to their overall character. Such districts are subject to rapid change, however. As more and more better-paid households are attracted to the creative buzz and lower property values of neo-bohemia, less affluent households are displaced to 'new bohemias' in alternate or adjacent transitional districts.

Distinctive design cultures can emerge as a result of the particular mix of local cultural and design-related activities and the local taste systems of producers and consumers, along with the general influences of specific national and regional cultures. Successful communities of design practice add to a city's overall status and can influence not only the city's brand image but also that of other products associated with the city. As talented designers are drawn from elsewhere, the buzz of clusters, the potency of branding and the momentum of cumulative causation are all reinforced. The history, character, composition and spatial organization of design services vary from one city to another, however, as illustrated by three of the most prominent world cities in terms of design: Paris, New York and Milan.

Further reading

Christopher Breward and David Gilbert (eds) (2006) *Fashion's World Cities*. Oxford: Berg. Examines the relationship between metropolitan modernity and fashion culture, looking at the significance of certain key sites in fashion's world order and at transformations in the connections between key cities.

Philip Cooke and Luciana Lazzeretti (eds) (2008) *Creative Cities, Cultural Clusters and Local Economic Development*. Cheltenham: Edward Elgar. Analyses the economic development of cities from the 'cultural economy' and 'creative industry' perspectives, examining and differentiating them as two related but distinct segments of contemporary city economies.

Elizabeth Currid (2007) *The Warhol Economy: How Fashion, Art and Music Drive New York City*. Princeton, NJ: Princeton University Press. Currid uses the dynamism and hybridity of Warhol's famous 'Factory' as a metaphor for the creative economy within New York City.

Donald McNeill (2009) *The Global Architect: Firms, Fame and Urban Form*. London: Routledge. Explores the influence of globalization processes on urban change, architectural practice and the built environment, emphasizing the 'star system' of international architects and the role of advanced information technology in expanding the geographical scope of the industry.

Robert O'Byrne (2009) *Style City: How London Became a Fashion Capital*. London: Frances Lincoln. Describes and illustrates the key players and influences of British fashion from the 1970s to 2000: not only the designers but also the music, the clubs and London itself.

Agnès Rocamora (2009) *Fashioning the City: Paris, Fashion and the Media*. London: I.B. Tauris. Focuses on how the French fashion press has been able to construct Paris as a leading world fashion city.

8 Conclusion

Toward liveability and sustainability

This concluding chapter looks at ideas about liveability and sustainability. Both have recently increased in political and policy importance. Drawing on a long history of ideas, concepts and experience, current thinking tends toward more holistic approaches that acknowledge the place of design within the broader political economy. 'True urbanism' recognizes the multiple actors in systems of provision as well as the need for flexibility and diversity in guiding urban development. 'Integral urbanism' recognizes the need for greater integration in the tasks that planners and architects typically conceive of as being separate from each other. Sustainability recognizes the interdependencies among issues affecting not only the environment but also social justice and the economy.

Beyond the fundamental importance of design in facilitating the circulation and accumulation of capital, design can also play key roles, as indicated at the beginning of this book, in contributing to the quality of life and the liveability and sustainability of cities. Design can make urban environments more legible and help to create a sense of place. It can help people with physical disabilities through codified 'universal' design standards. It can promote and ensure public health and bring order and stability to otherwise complex, chaotic and volatile settings. It can make transportation and land use more efficient. It can prevent crime, protect built heritage, engender community and encourage conviviality.

As we have seen, though, designers must operate within an economic and cultural environment that has for decades been driven by consumerism; they must work under what sociologist Magali Sarfatti-Larson (1993: 23) refers to as 'heteronomous conditions', in which practitioners are caught in a 'permanent contradiction' between, on the one hand, the autonomy of their own design ideals and solutions and, on the other, a matrix of relational networks and socio-institutional and project contexts involving interdisciplinary teams, and powerful clients, financiers, political actors and regulators. Design outcomes are always contingent and negotiated, therefore; while they may contribute more than 'artful fragments' to the city (Boyer 1988), the net result is that designers are significantly constrained in their ability to counter the inherent inequalities, socio-cultural conflicts and environmental problems of contemporary cities.

In a recent commentary, Leslie Sklair (2009) has posed the question of what might be done in terms of architecture, urban design and planning in order to move beyond artful fragments toward more progressive and 'emancipatory' cities. Consumerism, he argues, has become oppressive 'as it inevitably exacerbates the twin crises of capitalist globalization – namely, class polarization and ecological unsustainability' (Sklair 2009: 2704). After several decades of neoliberal political economy, the financial market crisis and economic recession of 2008–2009 has opened up the opportunity for the evolution of a new paradigm of policy and practice in urban design and planning, with a new (but as yet unnamed and tentatively shaped) regime of accumulation with a significant degree of reinstated government control and significant public investment in urban infrastructure (Knox and Schweitzer 2010). It has also brought the culture-ideology of consumerism into question among a broad public, thus raising the possibility, once again, of realizing the promise of design in creating more liveable, sustainable cities.

Toward 'true urbanism'

There is no shortage of ideas about liveability and sustainability among the design community. From any perspective, liveability is a complex, multifaceted concept. It is also a highly relative term: what is considered a 'liveable' community in one part of the world might be deemed highly unsatisfactory in another. Nevertheless, the idea of liveability remains a powerful one and, as we have seen, there has been a long history of engagement with the notion of liveability within the design professions. Since the mid-twentieth century the discourse has been framed largely around various reactions to the dominant paradigm of Modernist design. Lewis Mumford, observing the rapid industrialization of the 1920s followed Patrick Geddes in arguing for the preservation of regional architectural traditions. In the 1950s a 'Townscape Movement' emerged in Britain as a reaction to the sculptural architecture of Modernism and the lack of urbanity and human scale in the first British New Towns. The movement stressed the 'art of relationship' among elements of the urban landscape and the desirability of the 'recovery of place' through pictorial composition: unfolding sequences of street scenes, buildings that enclose intimate public spaces, and variety and idiosyncrasy in built form.

In 1960 Kevin Lynch, on the planning faculty at the Massachusetts Institute of Technology, introduced the notion of the 'legibility' of townscapes, noting that people's perceptions of the built environment are dominated by a few key elements: landmarks, paths, nodes, edges and districts. In the 1970s Christopher Alexander, an English architect who taught at Berkeley, sought to identify a 'language' of patterns among elements of built form and public spaces, his rationale being that a knowledge of such patterns might be useful in imbuing urban design with 'timeless' sensibilities (Alexander 1977, 1979). The work of Lynch and Alexander has been influential, largely because it lent an analytic dimension to a growing concern for the qualitative aspects of urban landscapes. As Edward Relph (2004) observes:

> [T]ownscapes are simultaneously the contexts of temporal experiences and subject to temporality. They are the settings for diurnal, weekly, and seasonal patterns of human activity, the backdrops and reference points for recollections and expectations. They are an essential component of the geography of memory. And in a manner broadly similar to that of human life, albeit at many different and overlapping tempos, landscapes have rhythms of creation, change, and decay.
>
> (Relph 2004: 113)

A rather different perspective on the liveability of the built environment came from the neorationalist movement, led by the Italian scholar-practitioner Aldo Rossi. His book *Architecture of the City* (1966) sought to identify various 'types' of architecture appropriate to economic and geographic context as an alternative to the totalizing models of Modernist architecture. Neorationalists saw the built environment as a 'theatre of memory' and hoped to identify

> the fundamental types of habitat: the street, the arcade, the square, the yard, the quarter, the colonnade, the avenue, the centre, the nucleus, the crown, the radius, the knot [. . .] So that the city can be walked through. So that it becomes a text again.
>
> (Robert Delevoy, quoted in Ellin 1996: 10)

During the 1970s these ideas were pursued by the Movement for the Reconstruction of the European City, with Léon Krier as its chief exponent. Krier was an advocate of urban design based on identifiable, functionally integrated quarters and of architecture with the proportions, morphology and craftsmanship of the pre-industrial era, with set-piece ensembles of buildings.

Another important contribution to ideas on liveability was the Museum of Modern Art's exhibition on 'Architecture Without Architects' in New York in 1964, which greatly stimulated an interest in vernacular architecture and sense of place (Rudofsky 1964). In France, the impulse to regenerate traditional urban qualities resulted in a movement in favour of 'Provincial Urbanism', and several developments in this vein gained international attention. One was Port Grimaud, designed and developed by François Spoerry near Saint Tropez in 1973 to resemble a fishing village. In Italy, the inaugural international architectural exhibition of the Venice Biennale in 1980 took the theme of 'The Presence of the Past: The End of Prohibition', seeking to recast urban design theory by 'reawakening the imaginary'. In the United States, 'contextualism', as advocated by architectural theorist Colin Rowe (1975), placed special emphasis on drawing upon all of the inherited elements of the built environment – the street, the axis and building mass – as a definer of urban space. Meanwhile Kenneth Frampton (1983) called for a more 'critical regionalism' that might assimilate genuine local materials, crafts, topographies and climate with the broader trends of national and global culture.

More recently, New Urbanism (Chapter 5) has enlivened interest in urban design and brought fresh ideas to what had become routinized and bureaucratized issues of land use, urban design and planning. It has also reinforced sense of place, liveability, sustainability and quality of life as important policy issues, and helped to resurrect the idea of a definable public interest. Unfortunately, its market success has been very much in the context of consumerism and neoliberalism, its progressive ideals watered down by developers exploiting the New Urbanism label for branding and marketing while only meeting a few of the principles of the movement's Charter.

'True urbanism', observes Seattle-based architect, city planner and urban designer Mark Hinshaw (2005: 26–27), is 'not the product of a singular vision' but, rather, emerges 'from the collective decisions of many organisations, associations, and government bodies'. True urbanist communities, he argues, 'are constantly evolving, infilling, and re-developing, with a broad mixture of architectural styles and sensibilities. [. . .] They have a gritty urbanity that values variety over uniformity'. The focus of urban design should be

> on the diversity and activity which help to create successful urban places, and, in particular, on how well the physical milieu supports the functions and activities taking place there. [. . .] With this concept comes the notion of urban design as the design and management of the 'public realm' – defined as the public face of buildings, the spaces between frontages, the activities taking place in and between these spaces, and the managing of these activities, all of which are affected by the uses of the buildings themselves.
>
> (Carmona *et al*. 2003: 7)

In Britain, research carried out on behalf of the Department for Communities and Local Government as part of a State of the English Cities project noted that the political and policy importance of the liveability of places has been rising, with the public placing a greater emphasis on local environmental quality than ever before. This research views liveability as a subset of quality of life, one that is concerned primarily with the quality of space and the built environment. Thus liveability 'is about how easy a place is to use and how safe it feels. It is about creating – and maintaining – a sense of place by creating an environment that is both inviting and enjoyable' (Department for Communities and Local Government 2006: 15).

According to Suzanne Lennard of International Making Cities Liveable orga-nization (IMCL), good urban design can not only 'enhance the well-being of inhabitants of towns', but also 'strengthen community, improve social and physical health, and increase civic engagement' (Lennard 2009). The Making Cities Liveable movement also promotes the idea of 'True Urbanism', with a perspective that draws heavily on the ideas of Geddes, Lynch, Alexander and Léon Krier. Thus True Urbanism, for IMCL, is based on 'time-tested principles' that emphasize the importance of the quality of public spaces (especially squares and marketplaces); of human-scale architecture with mixed-use structures that accommodate both retail and residential functions; of a compact urban fabric of blocks, streets and squares; and of outdoor cafés and restaurants, farmers' markets and community festivals. True Urbanism seeks to create 'places of short distances' where balanced transportation planning makes possible commuting via pedestrian networks, bicycle networks, traffic-quietened streets and public transportation. The Making Cities Liveable movement also places great emphasis on the inherited identity of towns – their 'DNA' – and seeks to promote public art and the idea of the built environment itself as a work of art.

Integral urbanism

Design's contribution to liveability, then, is not simply about form and morphology. It is about content, context and the capacity to foster conviviality, rhythm and movement. Liveable places should have plenty of opportunities for informal, casual meetings; street markets; a variety of comfortable places to sit, wait and people-watch; friendly 'third' places (cafés, pubs, bars, coffee houses and so on: Oldenburg 1999); and, above all, a sense of identity, belonging, authenticity and vitality. Architectural theorist Nan Ellin (2006) has expressed this in terms of what she calls 'Integral Urbanism'. The key attributes of integral urbanism, she suggests, are *hybridity, connectivity, porosity, authenticity* and *vulnerability*. Hybridity and connectivity depend on juxtaposition, simultaneity and the combination and linking of urban functions, connecting people and activities at key points of intensity and along thresholds between districts. Porosity depends on the visual and physical integration of the historic and the contemporary, of nature and the built environment, and of the social, cultural and physical dimensions of a town. Authenticity depends on both large-scale and small-scale interventions that are responsive to community needs and tastes and that are rooted in local climate, topography, history and culture. Vulnerability depends on a willingness on the part of urban planners and designers to relinquish control, to let things happen, and to allow for serendipity.

These qualities place a premium on process rather than outcome, and on the symbiotic relationships between people and places. The goal of integral urbanism is to ensure places that are 'in flow', where a city's physical settings and people's experiences of them are inseparable and reliant upon one another. Ellin (2006) observes:

> Encountering a place that is not in flow, the French typically remark that it lacks soul (*Il n'a pas d'âme*). Americans tend to say that it lacks character. Places that are in flow are characterized by the French as *animé* (animated, spirited, or soulful) and by Americans as lively.
>
> (Ellin 2006: 7)

Life between buildings

It is the capacity of the built environment to sustain animation, conviviality and sociability in public spaces that is perhaps the most important contribution to liveability. As sociologist Jan Gehl (1996: 131) put it: 'Life between buildings is both more relevant and more interesting to look at in the long run than are any combination of coloured concrete and staggered building forms.' It is, he argues, the social experience that is key to liveability. 'Compared with experiencing buildings and other inanimate objects, being among other people, who speak and move about, offers a wealth of stimulation' (Gehl 1996: 24). Gehl's point is that urban design can influence how many people use a city's public spaces, how long individual activities last, and which kinds of activities can flourish in different settings. In this context, he makes a distinction between necessary activities (such as shopping or going to work), optional activities (such as taking a stroll or stopping for a coffee at a pavement café) and social activities (such as chance encounters, gossiping, bantering, storytelling, joking, flirtation and serious conversation).

Better quality public spaces offer more options for social activities and are critical to the liveability of cities (Project for Public Spaces 2009). Because people are attracted to other people, stationary activities – standing around talking, sitting and people-watching or reading a newspaper, napping, sunbathing, sitting at a pavement café – are the ones that bring life to public spaces (Figure 8.1). In most settings, there is a marked 'edge effect' to these activities – people preferring to stay along the sides of streets and squares or in the transitional zones between one

Figure 8.1 Sheldon Square, London: part of the PaddingtonCentral redevelopment adjacent to the Grand Union Canal near Little Venice in central London, features a central space between the surrounding office buildings, shops and apartments. Although not purely public space, it provides a natural setting for informal social activities. (Photo: Stephen Cotton/ artofthestate)

space and the next, allowing them to linger and observe while remaining relatively inconspicuous. Familiarity, intersubjectivity and sociability tend to be reinforced so that positive affect (Chapter 6) is generated and social capital is developed.

With good public spaces, people's sense of civil society is intensified and the probability of their participation in local affairs and local democracy is increased. Squares and marketplaces are traditionally the loci of these activities, followed closely by pedestrianized streets and small parks. Yet public spaces are also spaces of circulation and movement, and there is also a need for balance between vehicle and pedestrian movement, and between optional movement and necessary circulation. Vehicle movements have a significant impact on the built environment and affect the quality of life for residents. This places a premium on urban design that facilitates and enhances pedestrian and, where topography and traffic density allows it, bicycle accessibility (Alfonzo *et al.* 2008; Leinberger 2008). In an increasingly fast-paced world, unhurried walking or cycling along safe and interesting routes is becoming a key dimension of liveability. Walking allows people to engage in a kind of mobile contemplation, a slow but thorough immersion in the rhythms of everyday life. Repetitive walking or bicycling along the same routes has the additional benefit of routine encounters with acquaintances and familiar faces, a precondition for familiarity, intersubjectivity, sociability and community.

Toward sustainable urbanism

As with liveability, there is a long tradition of advocacy of ideals and prescriptions in relation to the potential contributions of design to sustainability. Ian McHarg (1969), for example, was an early advocate of the idea that urban design should be based on ecological principles. Today, in the face of global warming and climate change and with widespread public concern over energy and water issues and environmental quality and greater consumer awareness about issues of health and well-being, there is renewed interest in how design can contribute to sustainability. Like liveability, sustainability is about the interdependent spheres of the economy, the environment and social well-being. This is often couched in terms of the 'Three Es of Sustainability' in urban development (Campbell 1996), referring to the environment, the economy and equity in society (Figure 8.2). It is a normative view that combines environmental sustainability with notions of economic growth and social justice. The difference between liveability and sustainability is that the latter involves a longer-term perspective. Sustainability means reviving economic growth; meeting essential needs for jobs, food, energy, water and sanitation; ensuring a balance between population and resources; conserving and enhancing the resource base; and managing risk, all without compromising – if possible – the ability of future generations to meet their needs.

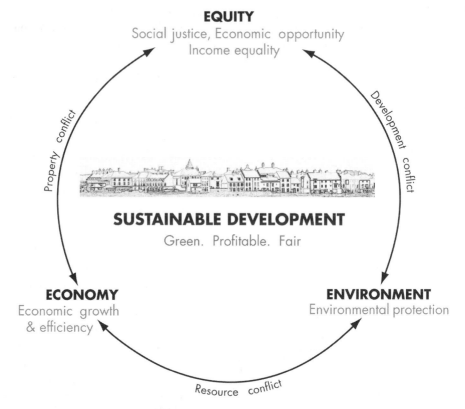

EQUITY
Social justice, Economic opportunity
Income equality

SUSTAINABLE DEVELOPMENT
Green. Profitable. Fair

ECONOMY
Economic growth
& efficiency

ENVIRONMENT
Environmental protection

Property conflict

Development conflict

Resource conflict

Figure 8.2 The Three Es of Sustainability. The circle illustrates the interdependencies between the three goals of sustainability. Considerations of equity, the environment and the economy can often stand in conflict with each other, but if they reinforce each other, they can lead to sustainable development. (Source: After S. Campbell (1996) 'Green Cities, Growing Cities, Just Cities? Urban Planning and the Contradictions of Sustainable Development', *Journal of the American Planning Association*, 62.3 and P.L. Knox and H. Mayer (2009) *Small Town Sustainability*. Basel, Birkhäuser, Figure 21.

Green design

As noted in Chapter 5, green design has emerged as a significant new below-the-line design movement. The idea of green design emerged from early critiques of consumerism such as Vance Packard's *The Waste Makers* (1960). In 1972, Victor Papanek wrote a scathing critique of conventional ideas of 'good design' in his book *Design for the Real World*:

Today, industrial design has put murder on a mass-production basis. By designing criminally unsafe automobiles that kill or maim nearly one million people around the world each year, by creating whole new species of permanent garbage to clutter up the landscape, and by choosing materials and processes that pollute the air we breathe, designers have become a dangerous breed.

(Papanek 1985: ix)

Papanek advocated socially and ecologically responsible design and it struck a sympathetic chord with a few practising designers who were looking for an alternative approach to design. Subsequently, his ideas have been carried forward and refined by authors such as Nigel Whiteley (1993) and Victor Margolin (1998), while there has been a significant upsurge in popular interest in ethical consumption.

In product design, green approaches today rest on the use of non-toxic, sustainably produced or recycled materials that require little energy to process; on longer-lasting and better-functioning products that will have to be replaced less frequently and that are designed for recycling; and on environmentally friendly packaging.

In architecture, green design begins with appropriate siting of buildings and involves the use of sustainably produced, reused or recycled building materials; alternative energy sources such as solar technology, geothermal heat pumps or wind power; rainwater harvesting for gardening and washing; and on-site waste management such as green roofs that filter and control stormwater runoff (Kibert 2007; Williams 2007). One of the best known buildings with a significant 'green' dimension is 30 St Mary Axe in London, also known as the Gherkin building (Figure 8.3), which uses half the energy a similar office tower would typically consume. Gaps around each floor plate create shafts that serve as a natural ventilation system for the entire building, creating a giant double-glazing effect. The shafts pull warm air out of the building during the summer and warm the building in the winter using passive solar heating.

An important part of the green design movement has been the development of design guidelines and codes. In the United Kingdom, for example, the Commission for Architecture and the Built Environment (CABE) has been producing design guidelines and directives since the early 2000s. In the United States, the US Green Building Council has developed Leadership in Energy and Environmental Design (LEED) standards for green buildings (Ben-Joseph 2009). Different levels of green building certification are awarded, based on a relatively straightforward checklist. As a result, architects and builders know that no matter where they are working, their buildings will be held to the same standards; similarly, clients and customers know just how 'green' their building will be. It is a voluntary system, but it is growing in popularity, partly as a result of market demand, and partly because it

Figure 8.3 The 'Gherkin', 30 St Mary Axe, London. The Lloyds Building is in the foreground. Formerly the SwissRe Building, 30 St Mary Axe is 180 metres (591 feet) tall, with forty floors. Its construction symbolized the start of a new high-rise construction boom in London. Designed by Norman Foster and Ken Shuttleworth, it was completed in 2004. (Photo: author)

offers developers a means of product differentiation for more affluent and environmentally conscious market segments. No building can be considered truly green, however, unless it is in a sustainable urban setting. This has prompted the Congress for the New Urbanism to partner with the Natural Resource Defense Council in the United States in proposing LEED standards for neighbourhood development (LEED-ND). The idea is to encourage development teams, planners, and local governments to construct sustainable, compact neighbourhoods. LEED-ND would allow neighbourhoods to be graded in terms of four categories: smart location and linkage; neighbourhood pattern and design; green infrastructure and buildings; and innovation and design process (CNU 2009; Garde 2009).

Most commercial clients, however, decline to pay for the additional up-front costs that a green building entails. It is not surprising, therefore, that nine of the top ten green projects recognized by the American Institute of Architects Committee on the Environment for 2008 were educational facilities funded by public agencies or philanthropic foundations. The exception, the Macallen Building, a 140-unit condominium building in Boston, Massachusetts, explicitly incorporated green design as a way of marketing a green lifestyle while at the same time increasing revenue from the project. The building features innovative technologies that save over 600,000 gallons of water annually while consuming 30 per cent less electricity than a conventional building.

Sustainable urbanism

From the perspective of cities and urban development, green design is just part of a broader movement toward sustainability. Sustainable urbanism involves compact, transit-oriented development, adaptive reuse, pedestrian- and bicycle-friendly settings, co-housing (typically, housing clustered around a pedestrian-only common street or courtyard, with a common house where residents can meet, eat together, or organize collective activities such as child care), landscaping that preserves and enhances wetlands and natural habitat, and the inclusion of ecological goals and criteria in governance and policy (Beatley 2000; Lafferty 2001).

A good example is Vauban, just outside Freiburg, Germany (Figure 8.4). It has been built as a sustainable model district of 5,000 inhabitants and 600 jobs on the site of a former military base. Connected to the town centre by a tramway, it also qualifies as a Transit-Oriented Development. All the houses are constructed to a low-energy consumption standard, with 100 units designed to an ultra-low energy 'Passivhaus' standard. Other buildings are heated by a combined heat and power station burning wood chips, while many of the buildings have solar collectors or photovoltaic cells (Schroepfer and Hee 2008).

Figure 8.4 Vauban, an eco-neighbourhood near Freiburg, Germany, with green roofs, solar panels and pedestrian-friendly streets. It is widely held to be one of the most important examples of green design and planning worldwide. (Photo: Frederick Florin/AFP/Getty Images)

Concerns over global warming and climate change have given particular prominence to eco-developments that strive for a carbon-neutral footprint. This presents enormous challenges in developing integrated planning and regulation for infrastructure and construction, but there are, nevertheless, many demonstration projects and committed communities. Ben-Joseph (2009) notes that there are at least forty different ecoprojects in China, the best known being Dongtan, an 86 square-kilometre masterplanned community built on wetland at the mouth of the Chang Jiang (Yangtze) river on the outskirts of Shanghai. More ambitious is Masdar, being built next to Abu Dhabi International Airport. The master plan has been designed by Foster + Partners and provides for a town of 50,000 residents. The city is intended to be completely carbon neutral with a zero-waste ecology and will use solar power and narrow shaded streets to reduce energy use. Cars will be banned in favour a personal rapid transit (PRT) system.

Less spectacular, but in many ways more impressive, is the Swedish network of eco-municipalities. The movement encompasses seventy-one municipalities of varying sizes – more than a quarter of the country's cities and towns (Knox and Mayer 2009). Each Swedish eko kommun works toward a sustainable future and collaborates with others. They have developed study circles in which residents can exchange ideas and develop a common understanding of what it means to develop a sustainable future. The idea for eco-municipalities originated in the 1980s when a small town close to the Arctic Circle, Övertoneå, motivated by a crisis in the form of economic and social decline, began to envision a fossil fuel-free future. Övertoneå became the first eco-municipality in Sweden and in the world. Today, Övertoneå has achieved 100 per cent independence from fossil fuels in its municipal operations. In 1995, Sweden's eco-municipalities created an umbrella organization known as Sveriges Ekokommuner. The organization has developed an environmental indicator system that municipalities can use to monitor progress toward sustainability (Sveriges Ekokommuner 2009). In the United Kingdom, meanwhile, the central government announced an eco-towns programme in 2009, involving new towns of 5,000–20,000 homes. The first towns in the programme include developments at Whitehill-Bordon, Hampshire; St Austell, Cornwall; Rackheath, Norfolk; and North West Bicester, Oxfordshire. They are intended to be exemplary green developments, and will be designed to meet the highest standards of sustainability, including low and zero carbon technologies and good public transport.

As if the challenge of achieving carbon-neutral footprints were not enough, we should bear in mind Scott Campbell's (1996) ideal of a balance among the 'Three Es'. Environmental sustainability must be achieved in conjunction with economic growth and efficiency, and equity, or social justice. In urban settings, the socio-economic dimensions of sustainability are critical. They include the need to maintain local socio-cultural attributes – neighbourliness and conviviality, for example – in the face of global influences and interdependencies. They also include aspects of social development that relate to the incidence of poverty and inequality, and accessibility to health care and education. Finally, they include aspects of social, cultural and political sensibilities that relate to a community's willingness and capacity to manage change in order to be more sustainable in a biophysical environmental as well as in an economic sense (Vallance 2007). The complexities and ambiguities involved in the interdependencies among the 'Three Es' within urban settings mean that the subject can be overwhelming, and for local planners and policy-makers this can lead to a kind of despairing inertia. Finding a balance between the economy, the environment and social well-being is not easy in practice because of various conflicts associated with the relationships between them. In particular, providing economic opportunities for a broad range of people almost inevitably conflicts with environmental protection. As Sklair (2009) acknowledges, much depends on society's willingness and ability to move away from the dominant culture-ideology of consumerism, toward a political economy that places greater emphasis on functionality, emancipation and sustainability.

Further reading

Timothy Beatley (2000) *Green Urbanism: Learning from European Cities*. Washington, DC: Island Press. Details the progress and policies of twenty-five of the most innovative cities in eleven European countries.

Scott Campbell (1996) 'Green Cities, Growing Cities, Just Cities? Urban Planning and the Contradictions of Sustainable Development', *Journal of the American Planning Association*, 62.3: 296–312. The seminal article on planning for urban sustainability.

Jan Gehl (2008) *Life Between Buildings: Using Public Space*, 6th edn. Copenhagen: Danish Architectural Press. A classic book about how public spaces are actually used, and how design can foster conviviality.

Nigel Whiteley (1993) *Design for Society*. London: Reaktion. Questions the role of design within consumer society and looks at the implications for design of the Green movement, the impact of feminism and the ideas of socially responsible designers.

References

Abu-Lughod, J. (1992) 'Disappearing Dichotomies', *Traditional Dwellings and Settlements Review*, 3: 7–12.

Alexander, C. (1977) *A Pattern Language: Towns, Buildings, Construction*. New York: Oxford University Press.

Alexander, C. (1979) *The Timeless Way of Building*. New York: Oxford University Press.

Alfonzo, M., Boarnet, M.G., Day, K., McMillan, T. and Anderson, C.L. (2008) 'The Relationship of Neighbourhood Built Environment Features and Adult Parents' Walking', *Journal of Urban Design*, 13.1: 29–51.

Amin, A. and Thrift, N. (2007) 'Cultural-Economy and Cities', *Progress in Human Geography*, 31.2: 143–161.

Anderson, B. (2006) 'Becoming and Being Hopeful: Towards a Theory of Affect', *Environment and Planning D, Society & Space*, 24: 733–752.

Arnold, D. and Hurst, W. (2004) 'End of the Iconic Age?', *Building Design*, 23 July (http://www.bdonline.co.uk/story.asp?sectioncode=426& storycode=3038558&c=1).

Aynsley, J. and Berry, F. (2005) 'Publishing the Modern Home: Magazines and the Domestic Interior 1870–1965', *Journal of Design History*, 18.1: 1–5.

Balzac, H. de (1837) *Illusions perdues*. Brussels: Nauman, Cattoir.

Banham, R. (1960) *Theory and Design in the First Machine Age*. London: Architectural Press.

Barnett, J. (1982) *An Introduction to Urban Design*. New York: Wiley.

Barthes, R. (1973) *Mythologies*, trans. A. Lavers. London: Paladin.

Basten, L. (2004) 'Perceptions of Urban Space in the Periphery: Potsdam's Kirchsteigfeld', *Tijdschrift voor Economische en Sociale Geografie*, 95: 89–99.

Bathelt, H., Malmberg, A. and Maskell, P. (2004) 'Clusters and Knowledge: Local Buzz, Global Pipelines and the Process of Knowledge Creation', *Progress in Human Geography*, 28.1: 31–56.

Baudelaire, C. (1965) *Art in Paris, 1845–1862*, trans. and edited J. Mayne. London: Phaidon.

Baudelaire, C. (1986) *The Painter of Modern Life and Other Essays*, trans. J. Mayne. New York: Da Capa. Orig. publ. 1859.

Baudrillard, J. (1968) *The System of Objects*. Bath: Bath Press.

Baudrillard, J. (1981) *For a Critique of the Political Economy of the Sign*. St Louis, MO: Telos.

Baudrillard, J. (1998) *The Consumer Society: Myths and Structures*. London: Sage.

Beatley, T. (2000) *Green Urbanism: Learning from European Cities*. Washington, DC: Island Press.

Beck, U. (1992) *Risk Society: Towards a New Modernity*. London: Sage.

Beck, U. (2006) *Power in the Global Age: A New Global Political Economy*. London: Polity.

Beck, U. (2008) *World at Risk*. London: Polity.

Beck, U. and Lau, C. (2005) 'Second Modernity as a Research Agenda: Theoretical and Empirical Explorations in the "Meta-change" of Modern Society', *British Journal of Sociology*, 56: 525–557.

Beck, U., Bonss, W. and Lau, C. (2003) 'The Theory of Reflexive Modernization', *Theory, Culture, and Society*, 20: 1–33.

Beck, U., Giddens, A. and Lash, S. (1994) *Reflexive Modernization: Politics, Tradition and Aesthetics in the Modern Social Order*. Stanford, CA: Stanford University Press.

Becker, H. (1984) *Art Worlds*. London: University of California Press.

Bell, D. (1976) *The Cultural Contradictions of Capitalism*. London: Heinemann Educational.

Bell, D. and Jayne, M. (2003) 'Assessing the Role of Design in Local and Regional Economies', *International Journal of Cultural Policy*, 9.3: 265–284.

Benjamin, W. (1999) *The Arcades Project*, trans. H. Eiland and K. McLaughlin. Cambridge, MA: Harvard University Press.

Ben-Joseph, E. (2009) 'Designing Codes: Trends in Cities, Planning and Development', *Urban Studies*, 46.12: 2691–2702.

Berman, M. (1983) *All That Is Solid Melts into Air*. London: Verso.

Beyerle, T. (2008) 'Giving Birth to Modernism', in Sudjic, D. (ed.) *Design Cities 1851–2008*. London: The Design Museum.

Biddulph, M. (2000) 'Villages Don't Make a City', *Journal of Urban Design*, 5: 65–82.

Bird, J. (1993) 'Dystopia on the Thames', in Bird, J., Curtis, B., Putnam, T., Robertson, G. and Tickner, L. (eds) *Mapping the Futures: Local Cultures, Global Change*. London: Routledge.

Blum, A. (2005) 'The Mall Goes Undercover', *Slate* (www.slate.com/id/2116246).

Blumer, H. (1969) 'From Class Differentiation to Collective Selection', *Sociological Quarterly*, 10.3: 275–291.

Bocock, R. (1993) *Consumption*. London: Routledge.

Bondi, L. (1991) 'Gender Divisions and Gentrification: A Critique', *Transcations of the Institute of British Geographers*, 16: 190–198.

Bondi, L. (1998) 'Gender, Class, and Urban Space: Public and Private Space in Contemporary Urban Landscapes', *Urban Geography*, 19.2: 160–185.

Bonta, J.P. (1979) *Architecture and its Interpretation*. London: Lund Humphries.

Booth, C. (1889) *Life and Labour of the People*, Volume I. London: Macmillan.

Borden, I. (2000) 'Fashioning the City', in Castle, H. (ed.) *Fashion + Architecture*. Chichester: Wiley-Academy.

Bottomore, T. (2003) *The Frankfurt School*. London: Routledge.

Bourdieu, P. (1984) *Distinction: A Social Critique of the Judgement of Taste*. London: Routledge & Kegan Paul.

Bourdieu, P. (1990) *In Other Words*. Cambridge: Polity.

Bourdieu, P. (1993) *The Field of Cultural Production*. New York: Columbia University Press.

Boyer, M.C. (1983) *Dreaming the Rational City*. Cambridge, MA: MIT Press.

Boyer, M.C. (1988) 'The Return of Aesthetics to City Planning', *Society*, 25.4: 49–56.

Branzi, A. (1994) *The Hot House: Italian New Wave Design*. London: Thames & Hudson.

Brenner, N. and Theodore, N. (2002) 'Cities and the Geography of "Actually Existing Neoliberalism" ', in Brenner, N. and Theodore, N. (eds) *Spaces of Neoliberalism: Urban Restructuring in North America and Western Europe*. Oxford: Blackwell.

Breward, C. and Gilbert, D. (eds) (2006) *Fashion's World Cities*. Oxford: Berg.

Bridge, G. (2006) 'Perspectives on Cultural Capital and the Neighbourhood', *Urban Studies*, 43: 719–730.

Brooks, D. (2004) *On Paradise Drive*. New York: Simon & Schuster.

Buchanan, R. (1995) 'Wicked Problems in Design Thinking', in Margolin, V. and Buchanan, R. (eds) *The Idea of Design*. Cambridge, MA: MIT Press.

Buck-Morss, S. (2002) *Dreamworld and Catastrophe: The Passing of Mass Utopia in East and West*. Cambridge, MA: MIT Press.

Bürger, P. (1984) *The Theory of the Avant-Garde*. Manchester: Manchester University Press.

Burton, E. and Mitchell, L. (eds) (2006) *Inclusive Urban Design: Streets for Life*. London: Architectural Press.

Butler, K. (1989) 'Pate Poverty: Downwardly Mobile Baby Boomers Lust after Luxury', *Utne Reader*, September/October: 75–82.

Butler, T. (2007) 'Re-urbanizing London Docklands: Gentrification, Suburbanization or New Urbanism?', *International Journal of Urban and Regional Research*, 31.4: 759–781.

Calthorpe, P. (1993) *The Next American Metropolis*. New York: Princeton Architectural Press.

Campbell, C. (1987) *The Romantic Ethic and the Spirit of Modern Consumerism*. Oxford: Blackwell.

Campbell, E. (2008) 'A Creative Centre', in Sudjic, D. (ed.) *Design Cities 1851–2008*. London: The Design Museum.

Campbell, S. (1996) 'Green Cities, Growing Cities, Just Cities? Urban Planning and the Contradictions of Sustainable Development', *Journal of the American Planning Association*, 62.3: 296–312.

Carmona, M. (2009a) 'The Isle of Dogs: Four Waves, Twelve Plans, 30+ Years, and a Renaissance . . . of Sorts', *Progress in Planning*, 71.3: 87–151.

Carmona, M. (2009b) 'Design Coding and the Creative, Market and Regulatory Tyrannies of Practice', *Urban Studies*, 46.12: 2643–2667.

Carmona, M., Heath, T., Oc, T. and Tiesdell, S. (2003) *Public Places, Urban Spaces: The Dimensions of Urban Design*. Oxford: Architectural Press.

Carter, E. (1979) 'Politics and Architecture: An Observer Looks Back at the 1930s', *Architectural Review*, November: 324.

Casellas, A. and Pallares-Barbera, M. (2005) 'Public-Sector Intervention in Embodying the New Economy in Inner Urban Areas: The Barcelona Experience', *Urban Studies*, 46.5–6: 1137–1155.

Caves, R. (2000) *Creative Industries: Contracts between Art and Commerce*. Cambridge, MA: Harvard University Press.

Chambers, J. (1992) *The Tyranny of Change: America in the Progressive Era, 1890–1920*. New Brunswick, NJ: Rutgers University Press.

Checkoway, B. (1980) 'Large Builders, Federal Housing Programs, and Postwar Suburbanization', *International Journal of Urban and Regional Research*, 4: 21–45.

Cinti, T. (2008) 'Cultural Clusters and Districts: The State of the Art', in Cooke, P. and Lazzeretti, L. (eds) *Creative Cities, Cultural Clusters and Local Economic Development*. Cheltenham: Edward Elgar.

CNU (Congress for the New Urbanism) (2009) 'LEED for Neighborhood Development' (www.cnu.org/leednd).

Cohen, L. (2003) *A Consumers' Republic: The Politics of Mass Consumption in Postwar America*. New York: Vintage.

Collins, M. and Papadakis, A. (1990) *Post Modern Design*. London: Academy Group.

Crewe, L. (1996) 'Material Culture: Embedded Firms, Organizational Networks and the Local Economic Development of a Fashion Quarter', *Regional Studies*, 30.3: 257–272.

Cross, N. (2001) 'Designerly Ways of Knowing: Design Discipline Versus Design Science', *Design Issues*, 17.3: 49–55.

Cuff, D. (1989) 'Through the Looking Glass: Seven New York Architects and their People', in Ellis, R. and Cuff, D. (eds) *Architects' People*. New York: Oxford University Press.

Currid, E. (2006) 'New York as a Global Creative Hub: A Competitive Analysis of Four Theories on World Cities', *Economic Development Quarterly*, 20.4: 330–350.

Currid, E. (2007) *The Warhol Economy: How Fashion, Art and Music Drive New York City*. Princeton, NJ: Princeton University Press.

Currid, E. and Connolly, J. (2008) 'Patterns of Knowledge: The Geography of Advanced Services and the Case of Art and Culture', *Annals of the Association of American Geographers*, 98.2: 414–434.

Cuthbert, A. (2006) *The Form of Cities: Political Economy and Urban Design*. Oxford: Blackwell.

Davis, M. (2006) *City of Quartz: Excavating the Future in Los Angeles*. London: Verso.

Dear, M. and Wolch, J. (1989) 'How Territory Shapes Social Life', in Dear, M. and Wolch, J. (eds) *The Power of Geography*. London: Unwin Hyman.

Debord, G. (1967) *La Société du spectacle*. Paris: Buchet-Chastel.

Debord, G. (1990) *Comments on the Society of the Spectacle*. London: Verso.

Debord, G. (1993) *The Society of the Spectacle*. New York: Zone.

de Certeau, M. (1984) *The Practice of Everyday Life*, trans. S. Randall. Berkeley, CA: University of California Press.

Defoe, D. (1927) *A Tour thro' the Whole Island of Great Britain, Divided into Circuits or Journies*. London: J.M. Dent.

de Franz, M. (2005) 'From Cultural Regeneration to Discursive Governance: Constructing the Flagship of the "Museumsquartier Vienna" as a Plural Symbol of Change', *International Journal of Urban and Regional Research*, 29.1: 50–66.

De Graaf, J., Wann, D. and Naylor, T.H. (2001) *Affluenza: The All-Consuming Epidemic*. San Francisco, CA: Berrett Koehler.

Department for Communities and Local Government (DCLG) (2006) *State of the English Cities: Liveability in English Cities*. London: DCLG.

Design Council (2005) *The Business of Design: Design Industry Research 2005*. London: The Design Council.

Deutsche, R. (1988) 'Uneven Development: Public Art in New York City', *October*, 47: 3–52.

Dexter, E. (2001) 'London 1990–2001', in Withers, R. (ed.) *Century City: Art and Culture in the Modern Metropolis*. London: Tate Gallery.

Domosh, M. (1996) 'The Feminized Retail Landscape: Gender Ideology and Consumer Culture in Nineteenth-century New York', in Wrigley, N. and Lowe, M. (eds) *Retailing, Capital and Consumption: Towards the New Retail Geography*. London: Longman.

Donald, S.H., Kofman, E. and Kevin, C. (eds) (2009) *Branding Cities: Cosmopolitanism, Parochialism, and Social Change*. London: Routledge.

Dormer, P. (1990) *The Meanings of Modern Design*. London: Thames & Hudson.

Dovey, K. (1999) *Framing Places: Mediating Power in Built Form*. London: Routledge.

Dovey, K. (2000) 'The Silent Complicity of Architecture', in Hillier, J. and Rooksby, E. (eds) *Habitus: A Sense of Place*. Aldershot: Ashgate.

Drake, G. (2003) ' "This Place Gives Me Space": Place and Creativity in the Creative Industries', *Geoforum*, 34: 511–524.

Duesenberry, J. (1949) *Income, Saving, and the Theory of Consumer Behavior*. Cambridge, MA: Harvard University Press.

Eco, U. (1986) *Travels in Hyperreality*. New York: Harcourt Brace Jovanovich.

Edwards, B. and Jenkins, P. (eds) (2005) *Edinburgh: The Making of a Capital City*. Edinburgh: Edinburgh University Press.

Ellin, N. (ed.) (1996) *Postmodern Urbanism*. Oxford: Blackwell.

Ellin, N. (2006) *Integral Urbanism*. New York: Routledge.

Ellis, C. (2002) 'The New Urbanism: Critiques and Rebuttals', *Journal of Urban Design*, 7.3: 261–291.

Entwistle, J. and Rocamora, A. (2006) 'The Field of Fashion Materialized: A Study of London Fashion Week', *Sociology*, 40.4: 735–751.

Evans, G. (2001) *Cultural Planning: An Urban Renaissance?* London: Routledge.

Evans, G. (2003) 'Hard-Branding the Cultural City – From Prado to Prada', *International Journal of Urban and Regional Research*, 27.2: 417–440.

Evans, G. (2005) 'Measure for Measure: Evaluating the Evidence of Culture's Contribution to Regeneration', *Urban Studies*, 42.5–6: 959–983.

Eyles, J. (1987) 'Housing Advertisements as Signs: Locality Creation and Meaning Systems', *Geografiska Annaler*, 69B: 93–105.

Featherstone, M. (1991) *Consumer Culture and Post-modernism*. London: Sage.

Fierro, A. (2006) *The Glass State: The Technology of the Spectacle*. Cambridge, MA: MIT Press.

Fine, B. and Leopold, E. (1993) *The World of Consumption*. London: Routledge.

Fishman, R. (1987) *Urban Utopias in the Twentieth Century*. Cambridge, MA: MIT Press.

Florida, R. (2002) *The Rise of the Creative Class: And How It's Transforming Work, Leisure, Community, and Everyday Life*. New York: Basic Books.

Florida, R. (2005) *Cities and the Creative Class*. New York: Routledge.

Florio, S. and Brownill, S. (2000) 'Whatever Happened to Criticism? Interpreting the London Docklands Development Corporation's Obituary', *City*, 4.1: 53–64.

Foot, J. (2001) *Milan Since the Miracle: City, Culture, and Identity*. Oxford: Berg.

Ford, L. (2008) 'World Cities and Global Change: Observations on Monumentality in Urban Design', *Eurasian Geography and Economics*, 49.3: 237–262.

Forty, A. (2005) *Objects of Desire: Design and Society Since 1750*. New York: Thames & Hudson.

Foucault, M. (1997) 'Space, Knowledge and Power', in Leach, N. (ed.) *Rethinking Architecture*. London: Routledge.

Fowler, B. and Wilson, F. (2004) 'Women Architects and their Discontents', *Sociology*, 38.1: 101–119.

Frampton, K. (1983) 'Prospects for a Critical Regionalism', *Perspecta*, 20: 147–162.

Frampton, K. (1991) 'Reflections on the Autonomy of Architecture: A Critique of Contemporary Production', in Ghirado, D. (ed.) *Out of Site: A Social Criticism of Architecture*. Seattle, WA: Bay Press.

Frampton, K. (1992) *Modern Architecture: A Critical History*. London: Thames & Hudson.

Frampton, K. (2007) *Modern Architecture: A Crirical History*, 4th edn. London: Thames & Hudson.

Frank, T. (1998) *The Conquest of Cool: Business Culture, Counter Culture and the Rise of Hip Consumerism*. Chicago, IL: University of Chicago Press.

Frisby, D. (2001) *Cityscapes of Modernity*. Cambridge: Polity.

Fuller, B. (1969) *Utopia or Oblivion*. Toronto: Bantam.

Galbraith, J.K. (1976) *The Affluent Society*. Boston, MA: Houghton Mifflin.

Garde, A. (2009) 'Sustainable by Design? Insight from U.S. LEED-ND Projects', *Journal of the American Planning Association*, 75.4: 424–440.

Garnham, N. (2005) 'From Cultural to Creative Industries: An Analysis of the Implications of the "Creative Industries" Approach to Arts and Media Policy Making in the United Kingdom', *International Journal of Cultural Policy*, 11.1: 15–30.

Garnier, T. (1918) *Une cité industrielle*. Paris: C. Massin.

Gasparina, J., O'Brien, G., Igarashi, T., Luna, I. and Steele, V. (2009) *Louis Vuitton: Art, Fashion and Architecture*. New York: Rizzoli.

Geddes, P. (1915) *Cities in Evolution*. London: Williams & Norgate.

Gehl, J. (1996) *Life Between Buildings: Using Public Space*. Copenhagen: Danish Architectural Press.

Ghirardo, D. (1984) 'Architecture of Deceit', *Perspecta*, 21: 110–115.

Gibson, D. (2009) *The Wayfinding Handbook: Information Design for Public Places*. Princeton, NJ: Princeton Architectural Press.

Giddens, A. (1979) *Central Problems in Social Theory*. London: Macmillan.

Giddens, A. (1984) *The Constitution of Society: Outline of the Theory of Structuration*. Cambridge: Polity.

Giddens, A. (1991) *Modernity and Self-identity*. London: Polity.

Gideon, S. (1948) *Mechanization Takes Command: A Contribution to Anonymous History*. New York: Oxford University Press.

Gilbert, D. (2006) 'From Paris to Shanghai: The Changing Geographies of Fashion's World Cities', in Breward, C. and Gilbert, D. (eds) *Fashion's World Cities*. Oxford: Berg.

Glass, R. (1968) 'Urban Sociology in Great Britain', in Pahl, R.E. (ed.) *Readings in Urban Sociology*. Oxford: Pergamon.

Gleeson, B. and Sipe, N. (eds) (2006) *Creating Child Friendly Cities: New Perspectives and Prospects*. London: Routledge.

Glennie, P. (1998) 'Consumption, Consumerism and Urban Form: Historical Perspectives', *Urban Studies*, 35.5–6: 927–951.

Gold, J.R. (2007) *The Practice of Modernism: Modern Architects and Urban Transformation, 1954–1972*. London: Routledge.

Goodman, D. and Chant, C. (eds) (1999) *European Cities and Technology*. London: Routledge.

Gospodini, A. (2002) 'European Cities in Competition and the New "Uses" of Urban Design', *Journal of Urban Design*, 7: 59–73.

Grabher, G. (2004) 'Learning in Projects, Remembering in Networks? Communality, Sociality, and Connectivity in Project Ecologies', *European Urban and Regional Studies*, 11.2: 103–123.

Granovetter, M. (1985) 'Economic Action and Social Structure: The Problem of Embeddedness', *American Journal of Sociology*, 91.3: 481–510.

Granovetter, M. (1991) 'The Social Construction of Economic Institutions', in Etzioni, A. and Lawrence, R. (eds) *Socio-economics: Towards a New Synthesis*. New York: Armonk.

Grant, J. (2006) *Planning the Good Community: New Urbanism in Theory and Practice*. New York: Routledge.

Greed, C. and Roberts, M. (eds) (1998) *Introducing Urban Design: Interventions and Responses*. London: Longman.

Greenberg, M. (2000) 'Branding Cities: A Social History of the Urban Lifestyle Magazine', *Urban Affairs Review*, 36.2: 228–263.

Greenberg, M. (2008) *Branding New York: How a City in Crisis Was Sold to the World*. London: Routledge.

Gregson, N., Crewe, L. and Brooks, K. (2002) 'Shopping, Space, and Practice', *Environment and Planning D: Society and Space*, 20: 597–617.

Gropius, W. (1926) 'Principles of Bauhaus Production', in Conrads, U. (ed.) *Programs and Manifestoes on Twentieth Century Architecture*. London: MIT Press.

Gutman, R. (1988) *Architectural Practice: A Critical Review*. New York: Princeton Architectural Press.

Gutman, R. (1989) 'Human Nature in Architectural Theory: The Example of Louis Kahn', in Ellis, R. and Cuff, D. (eds) *Architects' People*. New York: Oxford University Press.

Hackworth, J. (2007) *The Neoliberal City: Governance, Ideology, and Development in American Urbanism*. Ithaca, NY: Cornell University Press.

Hall, P. (1998) *Cities in Civilization*. London: Weidenfeld & Nicolson.

Hall, P. (2000) 'Creative Cities and Economic Development', *Urban Studies*, 37: 639–649.

Hall, P. (2002) *Cities of Tomorrow: An Intellectual History of Urban Planning and Design in the Twentieth Century*, 3rd edn. Oxford: Blackwell.

Harvey, D. (1985) *The Urbanization of Capital: Studies in the History and Theory of Capitalist Urbanization*. Baltimore, MD: Johns Hopkins University Press.

Harvey, D. (1989a) 'From Managerialism to Entrepreneurialism: The Transformation of Governance in Late Capitalism', *Geografiska Annaler*, 71(B): 3–17.

Harvey, D. (1989b) *The Condition of Postmodernity*. Oxford: Blackwell.

Harvey, D. (1998) 'Foreword', in Zukin, S. *Loft Living: Culture and Capital in Urban Change*. London: Century Hutchinson.

Harvey, D. (2000) *Spaces of Hope*. Berkeley, CA: University of California Press.

Harvey, D. (2003) *Paris, Capital of Modernity*. London: Routledge.

Hayden, D. (2003) *Building Suburbia: Green Fields and Urban Growth, 1820–2000*. New York: Pantheon.

Hebdige, R. (1979) *Subculture: The Meaning of Style*. London: Methuen.

Heidegger, M. (1971) *Poetry, Language, Thought*. New York: Harper & Row.

Herwig, O. (2008) *Universal Design: Solutions for Barrier-free Living*. Basel: Birkhäuser.

Heynen, H. (1999) *Architecture and Modernity: A Critique*. Cambridge, MA: MIT Press.

Hinshaw, M. (2005) 'The Case for True Urbanism', *Planning*, June: 24–27.

Hobbs, R., Lister, S., Hadfield, P., Winlow, S. and Hall, S. (2000) 'Receiving Shadows: Governance and Liminality in the Night Time Economy', *British Journal of Sociology*, 51: 701–717.

Hofseth, M. (2008) 'The New Opera House in Oslo – A Boost for Urban Development?', *Urban Research and Practice*, 1.1: 101–103.

Hollands, R. and Chatterton, P. (2003) 'Producing Nightlife in the New Urban Entertainment Economy: Corporatization, Branding and Market Segmentation', *International Journal of Urban and Regional Research*, 27.2: 361–385.

Howard, E. (1898) *To-morrow: A Peaceful Path to Real Reform*. London: Swann Sonnenschein.

Howard, E. (1902) *Garden Cities of To-morrow*. London: Swan Sonnenschein.

Hughes, R. (1980) *The Shock of the New: Art and the Century of Change*. New York: Knopf.

Hutton, T.A. (2004) 'The New Economy of the Inner City', *Cities*, 21.2: 89–108.

Imrie, R. and Street, E. (2009) 'Regulating Design: The Practices of Architecture, Governance and Control', *Urban Studies*, 46.12: 2507–2518.

Jackson, P. and Thrift, N. (1995) 'Geographies of Consumption', in Miller, D. (ed.) *Acknowledging Consumption*. London: Routledge.

Jacobs, J. (1958) 'Downtown is for People', in Whyte, W., Jr. (ed.) *The Exploding Metropolis*. Garden City, NY: Doubleday.

Jacobs, J. (1961) *The Death and Life of Great American Cities*. New York: Random House.

Jacobs, J. (1969) *The Economy of Cities*. New York: Random House.

James-Chakraborty, K. (2001) 'Kirchsteigfeld – A European Perspective on the Construction of Community', *Places*, 14: 56–63.

Jameson, F. and Fish, S. (1991) *Postmodernism, or, The Cultural Logic of Late Capitalism*. Durham, NC: Duke University Press.

Jayne, M. (2006) *Cities and Consumption*. London: Routledge.

Jencks, C. (2005) *The Iconic Building*. New York: Rizzoli.

Jencks, C. (2006) 'The Iconic Building is Here to Stay', *City*, 10.1: 3–20.

Jones, A. (2008) 'The Rise of Global Work', *Transactions, Institute of British Geographers*, 33: 12–26.

Jones, P. (2009) 'Putting Architecture in its Social Place: A Cultural Political Economy of Architecture', *Urban Studies*, 46.12: 2519–2536.

Jones, P. (2010) *The Sociology of Architecture*. Liverpool: Liverpool University Press.

Jordan, P.W. (2002) *Designing Pleasurable Products*. London: Taylor & Francis.

Jordan, P.W. (2007) 'The Dream Economy: Designing for Success in the 21st Century', *CoDesign*, 3, Supplement 1: 5–17.

Julier, G. (2000) *The Culture of Design*. London: Sage.

Julier, G. (2005) 'Urban Designscapes and the Production of Aesthetic Consent', *Urban Studies*, 42: 869–887.

Julier, G. (2006) 'From Visual Culture to Design Culture', *Design Issues*, 22.1: 64–76.

Katz, C. (1998) 'Excavating the Hidden City of Social Reproduction: A Commentary', *City and Society*, 10: 37–46.

Kelley, K.E. (2005) 'Architecture for Sale(s): An Unabashed Apologia', in Saunders, W.S. (ed.) *Commodification and Spectacle in Architecture*. Minneapolis, MN: University of Minnesota Press.

Kellner, D. (2005) 'Media Culture and the Triumph of the Spectacle', *Fast Capitalism*, 1.1 (www.uta.edu/huma/agger/fastcapitalism/1_1/kellner.html).

Kennedy, P. (2005) 'Joining, Constructing and Benefiting from the Global Workplace: Transnational Professionals in the Building-Design Industry', *Sociological Review*, 53.1: 172–197.

Kibert, C.J. (2007) *Sustainable Construction: Green Building Design and Delivery*, 2nd edn. Washington, DC: US Green Building Council.

Kim, J. and Kaplan, R. (2004) 'Physical and Psychological Factors in Sense of Community: New Urbanist Kentlands and nearby Orchard Village', *Environment and Behavior*, 36: 313–340.

Klingmann, A. (2007) *Brandscapes: Architecture in the Experience Economy*. Cambridge, MA: MIT Press.

Knox, P.L. (1984) 'Styles, Symbolism and Settings: The Built Environment and the Imperatives of Urbanised Capitalism', *Architecture et Comportement*, 2.2: 107–122.

Knox, P.L. (1987) 'The Social Production of the Built Environment: Architects, Architecture, and the Postmodern City', *Progress in Human Geography*, 21.3: 354–377.

Knox, P.L. (1991) 'The Restless Urban Landscape: Economic and Socio-Cultural Change and the Transformation of Washington, D.C.', *Annals, Association of American Geographers*, 81.2: 181–209.

Knox, P.L. (2008) *Metroburbia, USA*. New Brunswick, NJ: Rutgers University Press.

Knox, P.L. (2010) 'Starchitects, Starchitecture, and the Cultural Economy of Global Cities', in Derudder, B., Hoyler, M., Taylor, P.J. and Witlox, F. (eds) *International Handbook of Globalization and World Cities*. London: Edward Elgar.

Knox, P.L. and Cullen, J.D. (1981) 'Town Planning and Internal Survival Mechanisms of Urbanized Capitalism', *Area*, 13: 183–188.

Knox, P.L. and Mayer, H. (2009) *Small Town Sustainability*. Basel: Birkhäuser.

Knox, P.L. and Schweitzer, L. (2010) 'Design Determinism, Post-Meltdown: Urban Planners and the Search for Policy Relevance', *Housing Policy Debate*, 20.3: 267–277.

Knox, P.L. and Taylor, P.J. (eds) (1995) *World Cities in a World-System*. Cambridge: Cambridge University Press.

Knox, P.L. and Taylor, P.J. (2005) 'Toward a Geography of the Globalization of Architecture Office Networks', *Journal of Architectural Education*, 58.3: 23–32.

Knox, P.L., Agnew, J. and McCarthy, L. (2008) *The Geography of the World Economy*, 5th edn. London: Edward Arnold.

Kohn, M. (2004) *Brave New Neighborhoods: The Privatization of Public Space*. New York: Routledge.

Koskinen, I. (2005) 'Semiotic Neighborhoods', *Design Issues*, 21: 13–27.

Kraftl, P. and Adey, P. (2008) 'Architecture/Affect/Inhabitation: Geographies of Being-In Buildings', *Annals of the Association of American Geographers*, 98.1: 213–231.

Krier, R. and Kohl, C. (1999) *The Making of a Town: Potsdam Kirchsteigfeld*. London: Papadakis.

Lafferty, W.M. (ed.) (2001) *Sustainable Communities in Europe*. London: Earthscan.

Larner, W., Molloy, M. and Goodrum, A. (2007) 'Globalization, Cultural Economy, and Not-so-global Cities: The New Zealand Designer Fashion Industry', *Environment and Planning D: Society and Space*, 25: 381–400.

Lash, S. and Urry, J. (1994) *Economies of Signs and Space: After Organized Capitalism*. London: Sage.

Lasswell, H. (1979) *The Signature of Power*. New Brunswick, NJ: Transaction.

Law, A. and Mooney, G. (2005) 'Urban Landscapes', *International Socialism*, 106 (www.isj.org.uk/index.php4?id=95&issue=106).

Le Corbusier (1927) *Toward a New Architecture*. New York: Dover.

Lees, L. (2001) 'Towards a Critical Geography of Architecture: The Case of an Ersatz Colosseum', *Ecumene*, 8.1: 51–86.

Lees, L., Slater, T. and Wyly, E. (2008) *Gentrification*. London: Routledge.

Leinberger, C. (2008) *The Option of Urbanism*. Washington, DC: Island Press.

Lennard, S. (2009) 'Mission', International Making Cities Livable (www.livablecities.org/about/mission.html).

Leslie, D.A. (1993) 'Femininity, Post-Fordism, and the "New Traditionalism" ', *Environment and Planning D: Society and Space*, 11: 689–708.

Leslie, D.A. and Reimer, S. (2003a) 'Fashioning Furniture: Restructuring the Furniture Commodity Chain', *Area*, 35.4: 427–437.

Leslie, D.A. and Reimer, S. (2003b) 'Gender, Modern Design, and Home Consumption', *Environment and Planning D: Society and Space*, 21: 293–316.

Lévi-Strauss, C. (1970) *The Raw and the Cooked*. London: Cape.

Levy, J. (2008) *Contemporary Urban Planning*. Upper Saddle River, NJ: Prentice Hall.

Ley, D. (1996) *The New Middle Class and the Remaking of the Central City*. Oxford: Oxford University Press.

Ley, D. (2003) 'Artists, Aestheticisation and the Field of Gentrification', *Urban Studies*, 40: 2527–2544.

Lloyd, R. (2004) 'The Neighborhood in Cultural Production: Material and Symbolic Resources in the New Bohemia', *City and Community*, 3: 343–372.

Lloyd, R. (2005) *Neo-Bohemia: Art and Commerce in the Postindustrial City*. London: Routledge.

Lloyd, R. and Clark, T.N. (2001) 'The City as an Entertainment Machine', in Gotham, K.F. (ed.) *Critical Perspectives on Urban Redevelopment*. Amsterdam: JAI, Research in Urban Sociology.

Logan, J.R. and Molotch, H. (1987) *Urban Fortunes: The Political Economy of Place*. Berkeley, CA: University of California Press.

Lorentzen, A. and Hansen, C.J. (2007) 'Small Cities in the Experience Economy: An Evolutionary Approach', Regional Studies Association, Regions in Focus Conference, 2–5 April, Lisbon, Portugal.

Lorenzen, M. and Frederiksen, L. (2008) 'Why Do Cultural Industries Cluster? Localization, Urbanization, Products and Projects', in Cooke, P. and Lazzeretti, L. (eds) *Creative Cities, Cultural Clusters and Local Economic Development*. Cheltenham: Edward Elgar.

Low, S. and Lawrence-Zúñinga, D. (2003) 'Locating Culture', in Low, S. and Lawrence-Zúñinga, D., *The Anthropology of Space and Place*. Oxford: Blackwell.

Low, S. and Smith, N. (eds) (2005) *The Politics of Public Space*. London: Routledge.

Lynch, K. (1960) *Image of the City*. Cambridge, MA: MIT Press.

Lyotard, J-F. (1984) *The Postmodern Condition: A Report on Knowledge*. Minneapolis, MN: University of Minnesota Press.

McCracken, G. (1988) *Culture and Consumption: New Approaches to the Symbolic Character of Consumer Goods and Activities*. Bloomington, IN: Indiana University Press.

McHarg, I. (1969) *Design with Nature*. New York: Natural History Press.

McKenzie, E. (1994) *Privatopia: Homeowner Associations and the Rise of Residential Private Government*. New Haven, CT: Yale University Press.

Mackereth, S. (2000) 'Catwalk Architecture', in Castle, H. (ed.) *Fashion + Architecture*. Chichester: Wiley-Academy.

McNeill, D. (2000) 'McGuggenisation? National Identity and Globalisation in the Basque Country', *Political Geography*, 19: 473–494.

McNeill, D. (2009) *The Global Architect: Firms, Fame and Urban Form*. London: Routledge.

Madanipour, A. (2007) *Designing the City of Reason: Foundations and Frameworks in Urban Design Theory*. London: Routledge.

Maffesoli, M. (1996) *The Time of the Tribes: The Decline of Individualism in Mass Society*. London: Sage.

Margolin, V. (1998) 'Design for a Sustainable World', *Design Issues*, 14.2: 83–92.

Marin, L. (1990) *Utopics: The Semiological Play of Textual Spaces*. Amherst, NY: Prometheus.

Markusen, A. and Schrock, G. (2006) 'The Artistic Dividend: Urban Artistic Specialisation and Economic Development Implications', *Urban Studies*, 43.10: 1661–1686.

Marshall, A. (1890) *Principles of Economics*. New York: Macmillan.

Marshall, A. (2003) 'A Tale of Two Towns Tells a Lot about this Thing Called New Urbanism', *Built Environment*, 29: 227–237.

Martínez, J. M. (2007) 'Selling Avant-garde: How Antwerp Became a Fashion Capital (1990–2002)', *Urban Studies*, 44.12: 2449–2464.

Marx, L. (1964) *The Machine in the Garden*. New York: Oxford University Press.

Maskell, P. and Malmberg, A. (1999) 'Localised Learning and Industrial Competitiveness', *Cambridge Journal of Economics*, 23: 167–185.

Massey, D. (1991) 'Flexible Sexism', *Environment and Planning D: Society and Space*, 9: 31–57.

Massey, D. (2005) *For Space*. London: Sage.

Mearns, A. (1883) *The Bitter Cry of Outcast London: An Inquiry into the Condition of the Abject Poor* (pamphlet). London.

Merkel, J. (2000) 'Fashion Art in New York', in Castle, H. (ed.) *Fashion + Architecture*. Chichester: Wiley-Academy.

Miles, M. (2001) 'Picking Up Stones: Design Research and Urban Settlement', *Design Issues*, 17: 32–52.

Miles, M. (2005) 'Interruptions: Testing the Rhetoric of Culturally Led Urban Development', *Urban Studies*, 42.5–6: 889–911.

Miles, S. (1998) 'The Consuming Paradox: A New Research Agenda for Urban Consumption', *Urban Studies*, 35.5–6: 1001–1008.

Molotch, H. (2002) 'Place in Product', *International Journal of Urban and Regional Research*, 26: 665–688.

Mooney, G. (2004) 'Cultural Policy as Urban Transformation? Critical Reflections on Glasgow, European City of Culture 1990', *Local Economy*, 19.4: 327–340.

Morris, A. (1994) *History of Urban Form: Before the Industrial Revolutions*, 3rd edn. London: Longman.

Morris, W. (1882) *Hopes and Fears for Art: Five Lectures delivered in Birmingham, London and Nottingham*. London: Ellis and White.

Moudon, A.V. (2005) 'Active Living Research and the Urban Design, Planning, and Transportation Disciplines', *American Journal of Preventive Medicine*, 28.2: 214–215.

Mumford, E. (2000) *The CIAM Discourse on Urbanism, 1928–1960*. Cambridge, MA: MIT Press.

Mumford, L. (1938) *The Culture of Cities*. New York: Harcourt, Brace and World.

Naylor, G. (1968) *The Bauhaus*. London: Studio Vista.

Nelson, G. (1956) 'Obsolescence', *Industrial Design*, 6: 72–89.

Newman, O. (1972) *Defensible Space*. New York: Macmillan.

Nolen, J. (1927) *New Towns for Old: Achievements in Civic Improvement in Some American Small Towns and Neighborhoods*. Boston, MA: Marshall Jones.

O'Byrne, R. (2009) *Style City: How London Became a Fashion Capital*. London: Frances Lincoln.

Oldenburg, R. (1999) *The Great Good Place*. New York: Marlowe.

Packard, V. (1960) *The Waste Makers*. New York: D. McKay.

Pahl, R.E. (1969) 'Urban Social Theory and Research', *Environment and Planning A*, 1: 143–153.

Pallasmaa, J. (1996) *The Eyes of the Skin: Architecture and the Senses*. London: Academy Editions.

Papanek, V. (1985) *Design for the Real World: Human Ecology and Social Change*, 2nd edn. Chicago, IL: Academy Editions.

Paterson, M. (2006) *Consumption and Everyday Life*. London: Routledge.

Patton, P., Postrel, V. and Steele, V. (2004) *Glamour: Fashion, Industrial Design, Architecture*. New Haven, CT: Yale University Press.

Pawley, M. (2000) 'Fashion and Architecture in the 21st Century', in Castle, H. (ed.) *Fashion + Architecture*. Chichester: Wiley-Academy.

Peck, J. (2005) 'Struggling with the Creative Class', *International Journal of Urban and Regional Research*, 29: 740–770.

Peck, J. and Tickell, A. (2002) 'Neoliberalizing Space', in Brenner, N. and Theodore, N., (eds) *Spaces of Neoliberalism: Urban Restructuring in North America and Western Europe*. Oxford: Blackwell.

Pevsner, N. (1936) *Pioneers of the Modern Movement: From William Morris to Walter Gropius*. London: Harmondsworth.

Pinder, D. (2005) *Visions of the City*. New York: Routledge.

Pine, J. and Gilmore, J. (1999) *The Experience Economy*. Cambridge, MA: Harvard Business School Press.

Pirenne, H. (1952) *Medieval Cities: Their Origins and the Revival of Trade*. Princeton, NJ: Princeton University Press.

Postrel, V. (2003) *The Substance of Style: How the Rise of Aesthetic Value Is Remaking Commerce, Culture, and Consciousness*. New York: Perennial.

Power, D. and Hauge, A. (2008) 'No Man's Brand – Brands, Institutions, and Fashion', *Growth and Change*, 39: 123–143.

Poynor, R. (2003) *No More Rules: Graphic Design and Postmodernism*. London: Laurence King.

Poynor, R. (2007) *Obey the Giant: Life in an Image World*. Basel: Birkhäuser.

Pratt, A.C. (2000) 'New Media, the New Economy and New Spaces', *Geoforum*, 31: 425–436.

Pratt, A.C. (2008) 'Cultural Commodity Chains, Cultural Clusters, or Cultural Production Chains?', *Growth and Change*, 39.1: 95–103.

Pratt, A.C. (2009) 'Urban Regeneration: From the Arts "Feel Good" Factor to the Cultural Economy: A Case Study of Hoxton, London', *Urban Studies*, 46.5–6: 1041–1061.

Project for Public Spaces (2009) 'Our Mission'. Project for Public Spaces (www.pps.org/info/aboutpps/about).

Punter, J. (ed.) (2009) *Urban Design and the British Urban Renaissance*. London: Routledge.

Quinn, B. (2005) 'Arts Festivals and the City', *Urban Studies*, 42.5–6: 927–943.

Rantisi, N. (2002a) 'The Competitive Foundations of Localized Learning and Innovation: The Case of Women's Garment Production in New York City', *Economic Geography*, 78.4: 441–462.

Rantisi, N. (2002b) 'The Local Innovation System as a Source of "Variety": Openness and Adaptability in New York City's Garment District', *Regional Studies*, 36.6: 587–602.

Rantisi, N. (2004) 'The Designer in the City and the City in the Designer', in Power, D. and Scott, A.J. (eds) *Cultural Industries and the Production of Culture*. London: Routledge.

Ravetz, A. (1980) *Remaking Cities*. London: Croom Helm.

Reich, R. (1991) *The Work of Nations: Preparing Ourselves for 21st-Century Capitalism*. New York: Knopf.

Reich, R. (2007) *Supercapitalism: The Transformation of Business, Democracy, and Everyday Life*. New York: Knopf.

Reimer, S. and Leslie, D. (2008) 'Design, National Imaginaries, and the Home Furnishings Commodity Chain', *Growth and Change*, 39.1: 144–171.

Reimer, S., Pinch, S. and Sunley, P. (2008) 'Design Spaces: Agglomeration and Creativity in British Design Agencies', *Geografiska Annaler B*, 90: 151–172.

Reinach, S. (2006) 'Milan: The City of Prêt-à-porter in a World of Fast Fashion', in Breward, C. and Gilbert, D. (eds) *Fashion's World Cities*. Oxford: Berg.

Relph, E. (2004) 'Temporality and the Rhythms of Sustainable Landscapes', in Mels, T. (ed.) *Reanimating Places: A Geography of Rhythms*. Aldershot: Ashgate.

Ren, X. (2008) 'Architecture as Branding: Mega Project Developments in Beijing', *Built Environment*, 34.4: 517–531.

Rendell, J. (2000) 'Between Architecture, Fashion, and Identity', in Castle, H. (ed.) *Fashion + Architecture*. Chichester: Wiley-Academy.

Rishbeth, C. (2001) 'Ethnic Minority Groups and the Design of Public Open Space: An Inclusive Landscape?', *Landscape Research*, 26.4: 351–366.

Ritzer, G. (2005) *Enchanting a Disenchanted World: Continuity and Change in the Cathedrals of Consumption*, 2nd edn. Thousand Oaks, CA: Pine Forge Press.

Roach, J. (1996) *Cities of the Dead: Circum-Atlantic Performance*. New York: Columbia University Press.

Robbins, E. (2004) 'New Urbanism', in Robbins, E. and El-Khoury, R. (eds) *Shaping the City: Studies in History, Theory, and Urban Design*. New York: Routledge.

Roberts, G. and Steadman, P. (eds) (1999) *American Cities and Technology*. London: Routledge.

Rocamora, A. (2009) *Fashioning the City: Paris, Fashion and the Media*. London: I.B. Tauris.

Romanelli, M. (2008) 'Milan 1957', in Sudjic, D. (ed.) *Design Cities 1851–2008*. London: The Design Museum.

Rossi, A. (1966) *L'Architettura della Città*. Torino: Studi Edizioni.

Rothschild, J. (ed.) (1999) *Design and Feminism: Re-visioning Spaces, Places, and Everyday Things*. New Brunswick, NJ: Rutgers University Press.

Rowe, C. (1975) 'Collage City', *Architectural Review*, August: 65–91.

Rubin, B. (1979) 'Aesthetic Ideology and Urban Design', *Annals, Association of American Geographers*, 89: 339–361.

Rudofsky, B. (1964) *Architecture without Architects: An Introduction to Non-Pedigreed Architecture*. New York: Museum of Modern Art.

Ruskin, J. (1981) *The Stories of Venice*. Boston, MA: Faber and Faber. Orig. publ. 1851.

Ruskin, J. (1984) *Seven Lamps of Architecture*. New York: Farrar, Strauss and Giroux. Orig. publ. 1849.

Rybczynski, W. (2008) 'Architectural Branding', *The Appraisal Journal*, Summer: 279–284.

Sandercock, L. (1998) *Towards Cosmopolis: Planning for Multicultural Cities*. Chichester: Wiley.

Sarfatti-Larson, M. (1993) *Behind the Postmodern Facade: Architectural Change in Late Twentieth-century America*. Berkeley, CA: University of California Press.

Sassen, S. (2001) *The Global City*, 2nd edn. Princeton, NJ: Princeton University Press.

Sassen, S. (2002) *Global Networks, Linked Cities*. New York: Routledge.

Saunders, W.S. (2005) 'Preface', in Saunders, W.S. (ed.) *Commodification and Spectacle in Architecture*. Minneapolis, MN: University of Minnesota Press.

Saussure, F. de, Bouquet, S. and Engler, R. (2006) *Writings in General Linguistics*. Oxford: Oxford University Press. Orig. publ. 1916.

Schmitt, B. (1999) *Experiential Marketing*. New York: Free Press.

Schroepfer, T. and Hee, L. (2008) 'Emerging Forms of Sustainable Urbanism: Case Studies of Vauban Freiburg and solarCity Linz', *Journal of Green Building*, 3.2: 65–76.

Schwartz, J. (1998) 'Robert Moses', in Shumsky, N.L. (ed.) *Encyclopedia of Urban America*. Santa Barbara, CA: ABC-Clio.

Schwartz, V.R. (2001) 'Walter Benjamin for Historians', *American Historical Review*, 106.5: 1721–1743.

Scott, A.J. (2001) 'Capitalism, Cities, and the Production of Symbolic Forms', *Transactions of the Institute of British Geographers*, New Series, 26.1: 11–23.

Scott, A.J. (2004) 'Cultural-products Industries and Urban Economic Development. Prospects for Growth and Market Contestation in Global Context', *Urban Affairs Review*, 39: 461–490.

Scott, A.J. (2006) 'Creative Cities: Conceptual Issues and Policy Questions', *Journal of Urban Affairs*, 28: 1–17.

Scott, A.J. (2008) 'Resurgent Metropolis: Economy, Society and Urbanization in an Interconnected World', *International Journal of Urban and Regional Research*, 32.3: 548–564.

Scott, F.D. (2002) 'On Architecture under Capitalism', *Grey Room*, 6: 44–65.

Scully, V. (1988) *American Architecture and Urbanism*. New York: Henry Holt.

Sennett, R. (1997) 'The Search for a Place in the World', in Ellin, N. (ed.) *Architecture of Fear*. New York: Princeton Architectural Press.

Sherry, J. (1998) 'The Soul of the Company Store: Nike Town Chicago and the Emplaced Brandscape', in Sherry, J. (ed.) *ServiceScapes: The Concept of Place in Contemporary Markets*. Lincolnwood, IL: Ntc Business Books.

Short, J. and Kim, Y-H. (1999) *Globalization and the City*. London: Longman.

Simmel, G. (1903) *Die Grosstädte und das Geistesleben* [*The Metropolis and Mental Life*]. Dresden: Petermann.

Simmel, G. (1971) *On Individuality and Social Forms*. Chicago, IL: University of Chicago Press. Orig. publ. 1904.

Simms, A., Kjell, P. and Potts, R. (2005) *Clone Town Britain*. London: New Economics Foundation.

Simon, H. (1969) *The Sciences of the Artificial*. Cambridge, MA: MIT Press.

Sklair, L. (2005) 'The Transnational Capitalist Class and Contemporary Architecture in Globalizing Cities', *International Journal of Urban and Regional Research*, 29.3: 485–500.

Sklair, L. (2006) 'Iconic Architecture and Capitalist Globalization', *City*, 10.1: 21–47.

Sklair, L. (2009) 'Commentary: From the Consumerist/Oppressive City to the Functional/Emancipatory City', *Urban Studies*, 46.12: 2703–2711.

Smith, N. (1996) *The New Urban Frontier: Gentrification and the Revanchist City*. New York: Routledge.

Soja, E. (1980) 'The Socio-spatial Dialectic', *Annals of the Association of American Geographers*, 70: 207–225.

Soja, E. (2000) *Postmetropolis*. Oxford: Blackwell.

Sparke, P. (1990) '"A Home for Everybody?" Design, Ideology, and the Culture of the Home in Italy, 1945–72', in Greenhalgh, P. (ed.) *Modernism in Design*. London: Reaktion.

Stevens, G. (1998) *The Favored Circle: The Social Foundations of Architectural Distinction*. Cambridge, MA: MIT Press.

Sudjic, D. (2005) *The Edifice Complex: How the Rich and Powerful Shape the World*. London: Allen Lane.

Sudjic, D. (2008) 'Gathering a Head of Steam', in Sudjic, D. (ed.) *Design Cities 1851–2008*. London: The Design Museum.

Sunley, P., Pinch, S., Reimer, S. and Macmillen, J. (2008) 'Innovation in a Creative Production System: The Case of Design', *Journal of Economic Geography*, 8.5: 675–698.

Sveriges Ekokommuner (2009) 'Eco-municipalities' (http://sekom.sekom.nu/index.php?option=com_content&task=view&id=41&Itemid=50).

Swyngedouw, E., Moulaert, F. and Rodriguez, A. (2002) 'Neoliberal Urbanization in Europe: Large-Scale Urban Development Projects and the New Urban Policy', *Antipode*, 34.3: 543–577.

Tafuri, M. (1979) *Architecture and Utopia: Design and Capitalist Development*, trans. B. Luigia La Penta. Cambridge, MA: MIT Press.

Tafuri, M. and Co, F. (1986) *Modern Architecture 1*. New York: Rizzoli.

Tafuri, M. and Co, F. (1991) *Modern Architecture 2*. New York: Rizzoli.

Talen, E. (2008) *Design for Diversity: Exploring Socially Mixed Neighbourhoods*. London: Architectural Press.

Tallon, A. (2009) *Urban Regeneration in the UK*. London: Routledge.

Tarn, J. (1973) *Five Per Cent Philanthropy*. Cambridge: Cambridge University Press.

Taylor, P.J. (2004) *World City Network*. London: Routledge.

Taylor, P.J., Beaverstock, J., Cook, G. and Pandit, N. (2003) 'Financial Services Clustering and its Significance for London', Loughborough (GaWC Project 21: www.lboro.ac.uk/gawc/projects/projec21.html).

Thoreau, H.D. (1854) *Walden*. New York: T.Y. Crowell.

Thrift, N. (2004) 'Intensities of Feeling: Towards a Spatial Politics of Affect', *Geografiska Annaler B*, 86.1: 57–78.

Thrift, N. and Dewsbury, J-D. (2000) 'Dead Geographies – and How to Make Them Live', *Environment and Planning D: Society and Space*, 18: 411–432.

Tombesi, P. (2003) 'Super Market', *Harvard Design Magazine*, 17: 26–31.

Tombesi, P., Dave, B. and Scriver, P. (2003) 'Routine Production or Symbolic Analysis? India and the Globalization of Architectural Services', *Journal of Architecture*, 8: 63–94.

Tönnies, F. (1979) *Gemeinschaft und Gesellschaft: Grundbegriffe der reinen Soziologie*, 8th edn. Darmstadt: Wissenschaftliche Buchgemeinschaft. Orig. publ. 1887.

Turner, F.J. (1920) *The Frontier in American History*. New York: Henry Holt.

Twitchell, J. (2002) *Living It Up: America's Love Affair with Luxury*. New York: Simon & Schuster.

United States Bureau of Labor Statistics (2009) *Occupational Outlook Handbook, 2008–2009*. Washington, DC: Office of Occupational Statistics and Employment Projections.

Urban Task Force (1999) *Towards an Urban Renaissance: Report of the Urban Task Force*. London: Urban Task Force.

Vallance, S. (2007) 'The Sustainability Imperative and Urban New Zealand: Promise and Paradox', PhD dissertation, Lincoln University, New Zealand.

van Aalst, I. and Boogaarts, I. (2002) 'From Museum to Mass Entertainment: The Evolution of the Role of Museums in Cities', *European Urban and Regional Studies*, 9.3: 195–209.

Veblen, T. (1899) *The Theory of the Leisure Class*. New York: Macmillan.

Veninga, C. (2004) 'Spatial Prescriptions and Social Realities: New Urbanism and the Production of Northwest Landing', *Urban Geography*, 25: 458–482.

Venturi, R. (1966) *Complexity and Contradiction in Architecture*. New York: Museum of Modern Art.

Venturi, R., Brown, D.S. and Izenour, S. (1977) *Learning from Las Vegas: The Forgotten Symbolism of Architectural Form*. Cambridge, MA: MIT Press.

Vicario, L. and Monje, P.M. (2003) 'Another "Guggenheim Effect"? The Generation of a Potentially Gentrifiable Neighbourhood in Bilbao', *Urban Studies*, 40.12: 2383–2400.

Volkmann, C. and de Cock, C. (2006) 'Consuming the Bauhaus', *Consumption, Markets and Culture*, 9: 129–136.

Walker, J.A. (1989) *Design History and the History of Design*. London: Pluto.

Wansborough, M. and Mageean, A. (2000) 'The Role of Urban Design in Cultural Regeneration', *Journal of Urban Design*, 5.2: 181–197.

Warde, A. (1991) 'Gentrification as Consumption: Issues of Class and Gender', *Society and Space*, 9: 223–232.

Weber, N.F. (2008) *Le Corbusier: A Life*. New York: Knopf.

Weller, S. (2008) 'Beyond "Global Production Networks": Australian Fashion Week's Trans-Sectoral Synergies', *Growth and Change*, 39.1: 104–122.

Wenger, E. (1998) *Communities of Practice: Learning, Meaning, and Identity*. Cambridge: Cambridge University Press.

Whiteley, N. (1993) *Design for Society*. London: Reaktion.

Williams, D.E. (2007) *Sustainable Design: Ecology, Architecture, and Planning*. New York: Wiley.

Williams, R. (1973) *The City and the Country*. London: Chatto & Windus.

Wilson, E. (1991) *The Sphinx in the City: Urban Life, the Control of Disorder, and Women*. Berkeley, CA: University of California Press.

Wilson, E. (1995) 'The Rhetoric of Urban Space', *New Left Review*, 209: 146–160.

Wilson, E. (2003) *Adorned in Dreams: Fashion and Modernity*. New Brunswick, NJ: Rutgers University Press.

Wilson, E. (2006) 'Urbane Fashion', in Breward, C. and Gilbert, D. (eds) *Fashion's World Cities*. Oxford: Berg.

Wilson, W.H. (1989) *The City Beautiful Movement*. Baltimore, MD: Johns Hopkins University Press.

Wolfe, T. (1981) *From Bauhaus to Our House*. New York: Farrar Straus Giroux.

Wölfflin, H. (1966) *Renaissance and Baroque*, trans. K. Simon. Ithaca, NY: Cornell University Press.

Wright, P.H., Ashford, N.J. and Stammer, R.J. (1997) *Transportation Engineering: Planning and Design*. New York: Wiley.

Zola, E. (1867) *Thérèse Raquin*. Paris: Lacroix.

Zukin, S. (1991) *Landscapes of Power: From Detroit to Disneyworld*. Berkeley, CA: University of California Press.

Zukin, S. (1998) *Loft Living: Culture and Capital in Urban Change*. London: Century Hutchinson.

Zukin, S. (2004) *Point of Purchase*. London: Routledge.

Index

Page numbers in italics refer to illustrations